THE HUSH-KIT
BOOK OF
WARPLANES

THE HUSH-KIT BOOK OF
WARPLANES

EDITED BY JOE COLES

U

unbound

First published in 2022

Unbound
Level 1, Devonshire House, One Mayfair Place, London, W1J 8AJ
www.unbound.com

© Joe Coles, 2022
Foreword © Bill Sweetman, 2022
Contributions © David Axe, Dave 'Blo' Baranek, Bing Chandler, Joe Coles, Calum E. Douglas, Thomas Newdick,
Jim Smith, Ron Smith, Tim Robinson, Andreas Rupprecht, Kash Ryan, Edward Ward, Matt Willis, Pilot X, 2022
Fact-checked by Jon Lake
Chapters 22 and 24 translated from the Farsi by Kash Ryan
Illustrations by Inkworm, Penny Klein, Talgat Ashimov, the Teasel Studio
Diagrams by Made by Data

Text design by Amazing15

A CIP record for this book is available from the British Library

ISBN 978-1-80018-094-9 (hardback)
ISBN 978-1-80018-095-6 (ebook)

Printed in Slovenia by DZS Grafik

With thanks to:

Philip Rowles for suggesting 'A Banging History of the Bangseat'.

Nicholas Butta for suggesting 'The Unseen Menace: Interview with Lockheed F-117 Nighthawk pilot Major Robert "Robson" Donaldson'.

Trevor Beattie for suggesting Saro Lerwick in 'Top 10 Worst British Aircraft'.

Jim Kelly for suggesting 'Soviet Freak Show'.

Greg Cruz for suggesting 'A Brief History of Fighter Cockpits'.

And many others. You know who you are.

CONTENTS

FOREWORD

Joe Coles' Hush-Kit blog is aviation writing with a difference. Joe likes to find and tell the stories that others don't, or that just seldom get put into print. So, when he asked me to write a foreword, I thought I'd tell some stories of my own.

I was at the Paris Air Show in 1993 (I think), hanging out on the patio of a friendly chalet. There was a British guy there, one of a legion of plausible rogues seeking to generate peace, understanding and trade between the West and the ex-Soviet bloc, for a healthy commission. His associate, a Russian named Natasha, stood almost 6 feet tall and her role was to temporarily suppress the functional IQ of the Brit's potential business partners.

Enter the Harrier.

Natasha had doubtless seen many things but she had never seen a jet aeroplane halt in mid-air like a scruffy and extremely loud angel dancing on a wobbly pinhead, spin slowly around on its axis and drop a curtsy to the president of GIFAS (Groupement des Industries Françaises Aéronautiques et Spatiales). Muscovite reserve evaporated in an instant and she burst into laughter, clapping enthusiastically.

The love of warplanes is a deep paradox. They are designed to kill, injure and destroy, and they have done this on a larger scale than any other weapon. And yet even people who haven't been bitten by the aviation bug will stand mesmerised as a snorting, rumbling Second World War flyby rolls down the Potomac at rooftop height and over the National Mall; experience Emily Dickinson's 'zero at the bone' as the alien form of the B-2 sweeps over a football stadium; or simply watch, gobsmacked and half-deafened, as a fighter converts JP-8 into noise in a high-energy gymnastic performance.

The paradox is partly explained by the fact that warplanes are subtle killers – unlike, for example, a tank, with its Freudian steel tube waggling in your face. Like Mack the Knife, *das Messer sieht man nicht*, with bombs not being carried for fun and air-to-air missiles looking more like sporting projectiles than anything terribly lethal. What you see and hear is an exquisite machine designed to fly through the air and outperform other machines, whether in speed, range or agility – and in some cases to pass undetected by the most complex and powerful radars, or to land on 500 metres of road.

Aeroplane designers will seldom confess to being motivated by aesthetics. Styling is a bad word, and 'If it looks right, it'll fly right' is a cliché that the pros don't use. Engineers prefer the clunky and almost ugly phrase 'outer mould line' (OML) to

'shape' and like to imply that the OML is dictated by Newton, Bernoulli and sometimes Maxwell.

But if that was all there was to it, warplanes would all look the same, and they do not. The Spitfire and P-51 wings were designed to perform exactly the same mission (separated by four years of technology) but could not look more different from each other. A few more years separate the clearly F-4-related F-15 from the Su-27, but the Russian design is different, bolder and more spectacular.

At their best, aeroplanes are fascinating, three-dimensional sculptures, creations that make you stop, wonder and admire. I'll bet you that many people will get a bigger kick out of walking around an SR-71 than from staring at a Banksy monkey smoking a spliff. Certainly, I have taken many non-aviation people around the Steven F. Udvar-Hazy Center at Dulles International Airport, and the Blackbird always gets more attention than the space shuttle *Discovery*.

But like everything we can have opinions about and to which we can react emotionally, there is another level to the warplane bug. We humans are storytellers, and an important form of storytelling is mythology – and is there any Greco-Roman or Nordic deity whose name hasn't been applied to a warplane, missile or engine?

But there's a different level of mythology, as you can quickly find out by inputting certain key phrases online: Avro Arrow; TSR-2; 1957

OPPOSITE: The Avro Vulcan is a mess of beauty, destructive intent and national pride. High-speed aerodynamics, thunderous noise and world-class engineering combine to make military aircraft utterly seductive. Adding the headily narcotic appeal of national myth and apocalyptic potency results in a cult-like devotion to the Vulcan that will long outlive this Cold War bomber.

ABOVE: An RAF Harrier GR.3 unleashes four pods' worth of unguided rockets. Food photographers may worry about a mango looking unappetising, but this is relatively insignificant peril compared to the danger of shooting a fast-jet in flight launching a salvo of unpredictable unguided weapons.

RIGHT: Warplanes that were cancelled before they entered service have an allure all of their own. The remarkable Convair XFY Pogo was as impractical as it looks. Tail-sitters were one conceptual attempt at solving the riddle of combining vertical take-off and landing with a high top speed.

Defence White Paper. You'll more than likely incite impassioned debate about how superb aeroplanes were cancelled by weak, corrupt politicians and how much better things would have turned out for decades afterwards if that hadn't happened.

Now, to have that discussion in the 1960s and 1970s was one thing; Derek Wood's book *Project Cancelled* was a revelation in 1975, as were Bill Gunston's books and his articles in *Aeroplane Monthly*. But most of the people arguing the toss now weren't even born when all that was happening, and many are coming to the discussion primed by horrible robo-voiced YouTube videos. And so, myths are born.

Much of the myth-making is rooted in earlier myths, deliberately planted. Britain's aviation industry in the 1950s was coasting on its performance in the Second World War. 'Britain's lead in aviation' was a common phrase. My old *Observer's Book of Aircraft* credited the Supermarine Scimitar with supersonic speed. (Was any manufacturer such a one-hit wonder as Supermarine? Yes, Blackburn, but the retirement tour was a hit.) The industry wanted people to think that it was run by bold businessmen and staffed with brilliant engineers, yet held back by incompetent civil servants who could write impeccable Greek verse but had no idea how an aeroplane stayed up. Over long Fleet Street lunches its suave, ex-RAF PR men plied the myth with Homeric skill.

The truth was that most of the industry never lifted a finger without a government contract. Preserving historic names was more important than adapting the industry to lower, post-WW2 production rates, so small design teams with limited facilities ran late on their schedules. (By contrast, by 1960 France had pushed almost everyone except Dassault Aviation out of the warplane business.)

And the TSR-2? An elegant aircraft but deeply flawed. Someone had decided that it should be able to take off from a grass runway and nobody had the nerve to say otherwise. The engineers at Bristol knew that a three-point increase in Mach number would cost a fortune and incur a great increase in risk, but it was required anyway. The customer wasn't pinching pennies at all.

As for the avionics system, as described in now-declassified studies, it sounds horribly like the infamous Mark II system installed on the F-111D. The D-model was scrapped in favour of the simpler F-111E, and the ninety-six aeroplanes built were sent to Cannon Air Force Base in New Mexico, like a crazy aunt locked in the basement.

The Avro Arrow is a not-dissimilar story, except that the avionics system had barely been defined when the project was chopped. There was also the minor issue of the hordes of advanced Soviet bombers it was designed to destroy turning out not to exist.

So forgive me if I am short with people perpetuating these myths. The same goes for the Martin-Baker MB5, which may well have been a lovely aeroplane with a lot of sensible features, but during the Second World War Britain needed another fighter manufacturer like a hole in the head.

Is there a 'Greatest Warplane We Never Built'? As a journalist, I covered the story of one strong contender in the 1980s and 1990s: the F-16XL and the follow-on F-16U, which nearly got developed for the United Arab Emirates. The F-16's designer, Harry Hillaker, saw the delta-type wing as the way to go. He remarked once that you could always add power and external load but that eventually you ran out of wing area and space to put more stuff – a stretched fuselage and delta wing did both.

Much later, I gave a presentation on fighter developments to an audience that included Eurofighter people. One of them looked at the mid-nineties F-16U and said: 'It would have killed us.' Obviously, that did not happen. Why not? To quote Sherlock Holmes: 'It is associated with the giant rat of Sumatra, a story for which the world is not yet prepared.' One day. . .

Bill Sweetman, stealth-aircraft reporting guru and author of *Stealth Aircraft* and *Ultimate Fighter*

TOP: The Russian Su-35, one of the most advanced variants of the Flanker series. Despite its deployment in the villainous actions of Putin over Ukraine, it is hard not to admire its beastly good looks.

ABOVE: Strapping rockets to an aircraft, in this case an F-104, is one way to dramatically reduce take-off distance. As we will find out, the origins of this approach have a satanic influence.

TOP 10
FIGHTERS OF THE FIRST WORLD WAR

With a life expectancy of 18 flying hours, the first fighter pilots tended to have short and bloody existences. Terrifying and exhilarating in equal measure, human killing had found a new dimension. The fighter aircraft was invented in the First World War; the first machines designed explicitly to attack and destroy enemy aircraft appeared in 1915, and by 1918 thousands of fighters patrolled the skies. Here are the ten finest single-seat fighting scouts pilots took to war at the dawn of air-to-air combat.

10. Fokker Eindecker (1915)
'Fokker scourge!'

If the inspiration of terror were the sole criterion for a fighter's effectiveness, this spindly monoplane would be the most successful combat aircraft ever built. The Fokker was difficult to fly and its performance was pedestrian even for 1915, but it was equipped with the first (generally) reliable system allowing the gun to fire through the arc of the propeller. The pilot aimed the whole aircraft at the target and fired, a practice that remains standard for gun armament on fighter aircraft to this day. As soon as it was introduced, it was a fantastic success for the Germans, so much so that the planes became collectively known to the British as the 'Fokker scourge'. The first German aces accumulated victory tallies that the Allies could only be dream of. Max Immelmann, for example, downed a remarkable fifteen Allied aircraft solely with the Eindecker until his aircraft broke up in in mid-air – possibly because he shot off one of the plane's own propeller blades.

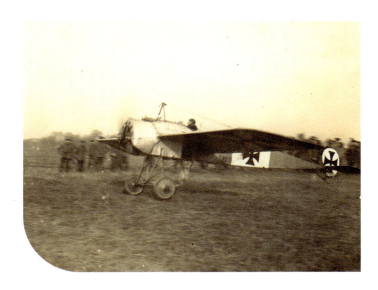

RIGHT: The famous forty-victory German ace Oswald Boelcke taking off in an Eindecker. The Fokker's greatest legacy was to ignite frenzied activity among the Allied designers to produce aircraft to defeat it, thus starting a see-sawing arms race of fighter aircraft that continued for the rest of the century.

'Navarre had painted his entire Nieuport scarlet and performed aerobatics to inspire the troops at Verdun'

9. Nieuport 17 (1916)
'The Scarlet Avenger'

During the First World War, France was the industrial powerhouse of aviation and the diminutive Nieuport Scout, as it was referred to, was the first truly mass-produced fighter, with at least 7,000 built in France. Compare this to the paltry 416 Fokker Eindeckers constructed and the licensed production of thousands more in Britain and Italy. The Germans paid the Nieuport the ultimate compliment by producing a direct copy that entered operational service as the Siemens-Schuckert D.I. The Nieuport lingered in frontline British service into early 1918 but was eventually surpassed by aircraft of greater power and strength, though some influential pilots retained one for personal use even when newer aircraft became available, notably Albert Ball and Charles Nungesser.

Probably the most interesting of the Nieuport's exponents was France's first official air ace, the mercurial 'Sentinel of Verdun', Jean Navarre. When Manfred von Richthofen, the future 'Red Baron', was still an obscure reconnaissance pilot, Navarre had painted his entire Nieuport scarlet and performed aerobatics to inspire the troops at Verdun, as well as over Amiens to impress women and, perhaps most audaciously, over the German aerodrome at Laon-Athies. 'Nobody thought to fire on me,' he explained. 'I

ABOVE: The Nieuport 17 was operated by seventeen nations during the First World War and for a time every French fighter squadron was equipped with one. Luckily, given its ubiquity, it was an outstanding aircraft and, initially at least, was superior to any enemy aircraft. This Soviet aircraft operated in the Russian Civil War and is actually a Nieuport 24bis, a development of the 17 with a slightly different wing but very similar performance.

was so proud.' His aerial antics were complemented by extreme levels of alcohol abuse, culminating in his drunkenly running over a gendarme with a Hispano-Suiza roadster after the unfortunate policeman had attempted to stop him driving on the pavement. The gendarme survived but Navarre was later killed in an accident while practising flying a Morane-Saulnier AI through the Arc de Triomphe for the 1919 Bastille Day parade.

8. Hanriot HD.1 (1916)
'Nungesser's Nunchaku'

Despite being rejected by the French, the Hanriot HD.1 was an important and effective aircraft which became the standard fighter of two major air arms. Hanriot were busy licence-building Sopwith 1½ Strutters when they designed the HD.1 and the Sopwith influence on the design is apparent. Intended as a replacement for the Nieuport 17, the Armée de l'air had decided to re-equip with the SPAD S.VII and declared the HD.1 inferior. A few went to the French Navy but Belgium, long used to receiving French cast-offs, jumped at the chance to operate a world-class fighter and the Hanriot was adopted as its standard fighter for the rest of the war, flying in the hands of many aces. Italy was even more enthusiastic, going so far as to build 831 examples of the Hanriot fighter under licence. Italian pilots rated the HD.1 a superior all-round fighter to the formidable SPAD S.XIII (which was also flown by Italy). For the massive offensive of October 1918, 130 of the 211 Italian fighter squadrons were equipped with the HD.1 and it flew in frontline Italian service until 1925.

In the same year the Hanriot made a bid for cinematic immortality in the *The Sky Raider*, a film starring forty-kill French ace Charles Nungesser (as himself) flying the HD.1 he had bought and shipped to the US to perform as part of a barnstorming 'flying circus'. Nungesser was approached by the film company who wished to cash in on his genuine air-ace credentials. As well as starring in the film, and flying a variety of the aircraft that appear on screen, Nungesser frequently appeared in person at screenings and performed aerobatics in the HD.1 over the cinema where the film was being shown.

ABOVE: Five-kill Hungarian ace Offiziersstellvertreter Friedrich Hefty flew this D.III, which is bedecked with elaborate swirls on all upper surfaces. Hefty was the first Austro-Hungarian pilot to escape by parachute when he was shot down in this aircraft in August 1918. He survived to serve in the Hungarian Air Force in the Second World War before emigrating to the United States.

BELOW: The Belgians were delighted with the HD.1. The Sopwith influence is plain to see, the HD.1 resembling a scaled-up Sopwith Pup. This snowbound Hanriot was operated by the 11ème Escadrille de Chasse in 1918.

7. Albatros D.III (Oeffag) (1916)
'Rois de l'azur'

The Albatros D.III and D.V development formed the bulk of the German fighter arm during 1917 and early 1918. The D.III proved formidable, being fast, well armed and generally sturdy, but it possessed a fundamental flaw: the lower wing was prone to failure in high-G-force manoeuvres. Strengthening improved the problem but it was never totally cured in German service. However, when Oesterreichische Flugzeugfabrik AG (Oeffag) in Austria started building the type under licence, their engineers altered the lower wing and eliminated the problem. Oeffag also dispensed with the large spinner of the German aircraft, which increased propeller efficiency and added 9 mph (14 km/h) to the maximum speed.

Austrian D.IIIs proved to be robust and effective aircraft that were popular with pilots and they remained in production until the Armistice. The model was probably the finest fighter available to the Central Powers until the introduction of the Fokker D.VII, and it remains a mystery why the successful modifications made by Oeffag were not applied to the German-built aircraft. After the war, further Oeffag D.IIIs were produced for Poland, with whom the aircraft saw action again during the Polish–Soviet War of 1919–1920.

6. Sopwith Camel 2F.1 (1916)
'The camel toe-truck'

The Camel shouldn't be omitted from any list of the 'greatest' First World War fighters – Camels accounted for 1,294 victories, more than any other fighter type of the war – but it was in many ways an extremely bad aircraft. Obsolete by the Armistice and already being relegated to ground-attack, it was famously difficult to fly as it had been designed to be unstable (manoeuvrability, like creativity, is aided by instability). Every major component of the aircraft (engine, fuel tank, armament and pilot) was crammed into the first 7 feet of the airframe and the aircraft was intended to be spectacularly manoeuvrable, which indeed it was. The torque of the rotary engine, some 375 lb (170 kg) spinning at the same speed as the propeller, meant it was constantly attempting to roll the aircraft in the opposite direction to its spin, and as a result the Camel turned right at an unbelievable speed – it may well have been the most manoeuvrable fighter of all time and was unbeatable in a turning fight. However, at low speed, the torque of the rotary engine was nearly beyond the ability of its rudder to correct, which caused frequent accidents on take-off or landing, with at least 385 Camel pilots losing their lives in non-combat-related

accidents. Pilots joked that the Camel offered the choice of a 'wooden cross, the Red Cross or a Victoria Cross'.

As well as the massive torque generated by a rotary engine and its effect on the handling of the aircraft, these motors could only be effectively lubricated by castor oil. Castor oil is a potent laxative and quantities of it were expelled by the spinning engine, some of which found its way to the pilot, leading to tales of daring young men flying into action while sitting in a pile of their own shit. This is largely apocryphal, however, not least as many rotary-engine aircraft are flying today and still lubricated by castor oil, and no pilot from 1918 to the present has yet confessed to rotary-induced incontinence. However, rotary engines also expel large quantities of unburned fuel to be inhaled by the pilot, and at least one modern pilot has admitted to feeling deeply nauseous an hour or two after flying such an aircraft. In the deadly world of air combat in the First World War, it does not seem ridiculous to suggest that the effects of breathing in fuel and castor oil for hours, while dealing with the mental turmoil of the prospect of sudden horrific death at any time, might have had a less than salutary effect on the digestive system.

ABOVE: The Camel's rival, the SE5a, was in virtually every respect a better aircraft, but the Camel seemed to capture the wider imagination, possibly due to the *Biggles* stories of Captain W. E. Johns and, of course, the type's most famous 'pilot', the cartoon character Snoopy. This image shows the Shuttleworth Collection's accurate replica, fitted with a genuine rotary engine.

5. Royal Aircraft Factory SE.5a (1916)
'Mannock's Maniac'

ABOVE: Although it possessed only adequate manoeuvrability, the SE5a was one of the fastest aircraft of the conflict, with top-ranking British ace James McCudden claiming: 'It was very fine to be in a machine that was faster than the Hun's, and to know that one could run away just as things got too hot.' Shown here is the Shuttleworth Collection's Wolseley-built example, which shot down a Fokker D.VII on the penultimate day of the war.

Representing the antithesis of the Camel, the SE5a complemented the Sopwith fighter to a remarkable degree. With its V8 Wolseley Viper engine, performance at altitude was excellent, giving it the advantage of height over its adversaries, at least until the advent of the Fokker D.VII, though it was never outclassed by the vaunted German fighter. Furthermore, the SE5a boasted forgiving handling, which the Sopwith Camel pilot could only dream of, offered very good visibility thanks to its unusually narrow fuselage and possessed an immensely strong airframe, imbuing it with greater survivability and crashworthiness than its peers. Flown by many aces, including Edward Mannock VC, the most successful British fighter pilot of all time, the SE5a was the best all-round British fighter available in quantity during the First World War. So why did an improved version never appear, in the same way that Sopwith's Camel gave rise to the Snipe? The answer is due, at least in part, to the activities of Noel Pemberton Billing, most famous today for founding the Supermarine company, which would later produce the Spitfire. An aviation-enthusiast MP with extreme right-wing views, Billing detested the Royal Aircraft Factory and

mounted a prolonged attack on its designs both in Parliament and the press. His idea, that the Royal Aircraft Factory was producing dangerous designs, primarily the BE2 series (described as 'Fokker Fodder'), was exaggerated but it gained much support during the war and ultimately contributed to the closure of the Factory's design office in 1919.

Both the credibility and persuasiveness of Billing can be surmised from his wartime claim that Germany was blackmailing '47,000 highly placed British perverts' to 'propagate evils which all decent men thought had perished in Sodom and Lesbia'. Their names were said to be inscribed in the Berlin 'Black Book', which purportedly showed that the Germans planned on 'exterminating the manhood of Britain' by luring them into homosexuality. In an article, 'The Cult of the Clitoris', he implied that the actress Maud Allan was a lesbian associate of the German conspirators. This led to a sensational libel case, at which Billing represented himself and won, despite the whole thing being very obviously a fabrication, a fact which Billing later admitted to having known all along.

4. SPAD S.XIII (1917)
'Fonck's Enforcer'

The SPAD S.VII combined a tough, compact airframe with the revolutionary V8 Hispano-Suiza engine, and its successor, the S.XIII, enhanced that combination to great effect. The new Hispano-Suiza 8Ba offered 200 hp and the aircraft's climb-and-dive performance was spectacular. In level flight the SPAD was roughly as fast as the SE5a, but it could outclimb the British aircraft. By contrast, its manoeuvrability was never considered anything more than adequate. As such, this most effective of French fighters ushered in a new form of aerial combat where performance in the vertical plane was paramount, and the sort of sparkling agility available with the Sopwith Camel became increasingly irrelevant.

Committed to combat in May 1917, French industrial might saw 8,472 SPAD S.XIIIs built during the war, with orders for around 10,000 more cancelled at the Armistice. By November 1918, every single-seat fighter in French service was a SPAD and the aircraft flew in Italian, Belgian, British and American units. A curious aspect of the SPAD story is that the man who founded the company that built the aircraft was in jail while it achieved its most spectacular success. Armand Deperdussin had founded the Société de Production des Aéroplanes Deperdussin in 1910, hiring ace designer Louis Béchereau. Unfortunately, he also embezzled 32 million Francs from his own firm, and in 1917 was jailed for five years, during which the company was renamed the Société Pour L'Aviation et ses Dérivés. Despite his major contribution to the dominance of French military aviation up to 1918, Deperdussin committed suicide in 1924, deciding to die rather than live in the misery of his fallen glory.

3. Fokker D.VII (1918)
'The Harbinger'

Often named the best fighter of the First World War, the Fokker D.VII wasn't the fastest, most manoeuvrable or best-climbing aircraft to see service but it offered the finest combination of these attributes in one highly robust airframe. Possessing a fine altitude performance, it was also easy to fly, with a particularly gentle stall, and could be flown effectively by inexperienced pilots. A massive production programme meant that, despite not entering service until May 1918, some 3,300 had been built by the end of hostilities that November. Curiously, both wings of the Fokker were cantilever units, virtually unheard of at the time, and as such the D.VII was a harbinger of a future where all fighters would have self-supporting wings. Wing struts were added to the design, but only to prevent vibration.

A curious footnote to the D.VII saga is the Fokker 'Ontwerp' (Design) 203 of 1940 (the Ontwerp 203 Sportflugzeug Projekt 1940). In a quite staggering example of toadying, directly after the Germans occupied the Netherlands, Friedrich Wilhelm Seekatz, manager of Fokker, decided to build a modern D.VII as a gift for Luftwaffe chief Hermann Goering, who had flown the type in 1918. Featuring a modern Argus engine in place of the original Mercedes unit, the 'new' D.VII was to feature a wider fuselage to allow for Goering's increased girth. Somewhat sadly, though undoubtedly correctly from a moral standpoint, the design was abandoned.

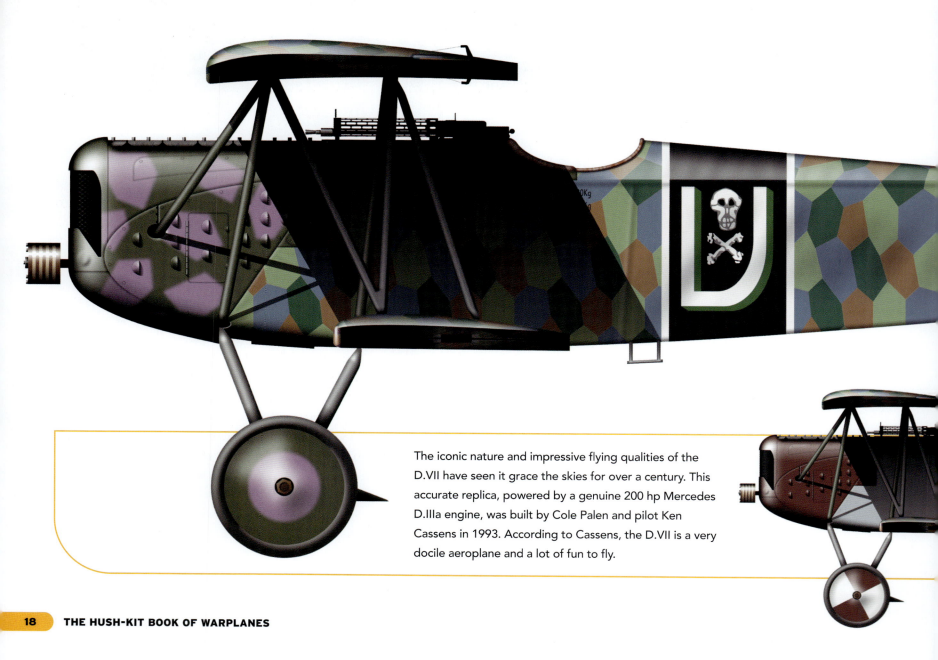

The iconic nature and impressive flying qualities of the D.VII have seen it grace the skies for over a century. This accurate replica, powered by a genuine 200 hp Mercedes D.IIIa engine, was built by Cole Palen and pilot Ken Cassens in 1993. According to Cassens, the D.VII is a very docile aeroplane and a lot of fun to fly.

RIGHT: Seen here with twin upward-firing Lewis guns, the Dolphin is an awkward-looking aircraft, but nonetheless has fair claim to being the finest British fighter of the First World War. The French-built Mk.II variant, produced to meet the needs of the Armée de l'air and US Army Air Service, was to be powered by a 300 hp Hispano-Suiza 8F, conferring an impressive 140 mph (225 km/h) top speed, but few were actually completed.

BELOW: Famously singled out to be handed over to the Allies as a condition of the Armistice, production of the D.VII ended in Germany in 1918. The D.VII re-entered production in the Netherlands and ultimately served in the air arms of nineteen nations, in some cases well into the 1930s. The example in this picture was flown by Erich Mix, who claimed three kills in 1918 and a further eight in the Second World War flying the Messerschmitt Bf 109, the last being in 1941, by which time he was forty-three years old.

2. Sopwith 5F.1 Dolphin
'Gillet's Killer'

Fitted with the same Hispano-Suiza 8B as the SPAD S.XIII, the Dolphin's entry into service was delayed by the scarce supply of engines – of the 2,072 Dolphins built, around 1,500 ended up in storage, unable to fly. Only four squadrons would ever wholly equip with the type. A measure of the aircraft's quality is the fact that France had selected a licence-built Dolphin to replace the formidable SPAD S.XIII, though only around twenty had been built by the Armistice.

During the relatively brief time it saw action, the Dolphin proved to be an outstanding fighter: fast, tough, manoeuvrable and often armed with four machine guns – double the standard armament of the time. Given its relative rarity, the Dolphin produced a remarkable number of aces; most successful was Francis Gillet, with 20 victories. Cecil Lewis, author of *Sagittarius Rising* (a masterful autobiographical account of aerial fighting lauded by many, including Bernard Shaw), described mock combat with a Dolphin in his SE5a: 'The Dolphin had a better performance than I realised . . . I sat in a tight climbing spiral, he sat in a tighter one. I tried to climb above him, he climbed faster. Every dodge I have ever learned I tried on him; but he just sat there on my tail, for all the world as if I had just been towing him behind me.'

'During the relatively brief time it saw action, the Dolphin proved to be an outstanding fighter'

'Boasting a colossal rate of climb, at height the D.IV could outrun and outmanoeuvre the Fokker D.VII, itself noted for its good high-altitude performance'

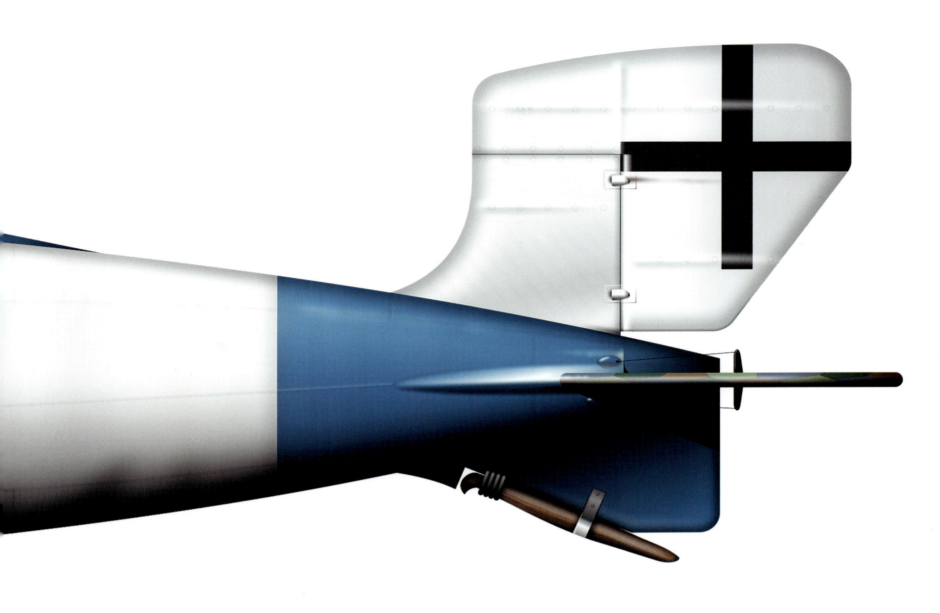

1. Siemens-Schuckert D.IV (1918)
'The Beast'

Seemingly something of a throwback, with its rotund, abbreviated fuselage and huge rotary engine, the Siemens-Schuckert D.IV was the finest interceptor to see service with Germany. Rotaries had reached the end of their development potential and the eleven-cylinder Siemens-Halske Sh.III fitted to the D.IV represented the zenith of this engine type, selected no doubt because it was a product of the same parent company as the airframe. By dint of an ingenious crank and gearing system, the torque that proved so deadly on other rotary powered aircraft, such as the Camel, was virtually eliminated, and a high compression ratio allowed the Sh.III to maintain an impressively high power output at altitude. Boasting a colossal rate of climb, at height the D.IV could outrun and outmanoeuvre the Fokker D.VII, itself noted for its good high-altitude performance, yet the D.IV could fly to 26,600 feet (8,100 m), about 3,900 feet (1,200 m) higher than the Fokker. As it appeared so late in the war, the excellent D.IV had little effect on the course of the conflict, and only 123 were built. Surprisingly, development continued after the cessation of hostilities, with Siemens-Schuckert producing a monoplane derivative in 1919, but the Treaty of Versailles rendered the development of further German military aircraft illegal.

ABOVE: Very few examples of the D.IV reached the front; fewer still saw combat. This example was flown by twenty-three-kill ace Hermann Becker, commander of Jagdstaffel 12, and used by him to claim his final two victories, both SPADs, in the final fortnight of the conflict.

THE ULTIMATE
BIPLANE
FIGHTERS

The biplane fighter had a mere twenty-year reign. During this period, many fine aircraft appeared, flew for a few years and were quietly withdrawn without ever firing a shot in anger. This list of the greatest examines only those 'lucky' enough to have seen operational combat service, before the monoplane showed two fingers to two wings for good. By the mid-1930s, the slight advantage in roll-rates and manoeuvrability at low speeds that the biplane offered were not worth the extra drag, reduced pilot view and additional complexity engendered by a second wing. To make the most out of new, powerful engines, a sleeker, faster type of fighter would come to dominate – the monoplane would now be king.

Polikarpov I-15 (1933)
'Soviet snubnose'

TOP SCORER: Leopoldo Morquillas Rubio – 21 kills, all flying the I-15

Compared to the decidedly elegant CR.32, the I-15 looks like it has been driven into a wall. But looks can be deceiving: the gull-winged Polikarpov was one of the few aircraft in the world capable of meeting the CR.32 on roughly even terms. Though a bit slower, the pugnacious I-15 was slightly more manoeuvrable; however, if neither aircraft had a height advantage, individual combat would ultimately be down to piloting skill. By contrast, the Germans' best fighter of the time, the Heinkel He 51, was completely outclassed by the Polikarpov, though they caught up quite quickly. Designed by Nikolai Polikarpov in ebullient mood (he had just been released from jail), the I-15 was conventional but brought Soviet fighter design up to contemporary world standards, where it remained until the Soviet Union's ultimate demise.

Weirdly nicknamed 'Curtiss' by the Nationalists, for reasons unclear to anyone, and more appropriately 'Chato' (a Spanish nickname for someone with a 'flat nose') by the Republicans who operated it, the I-15 was the premier fighter of the Republicans until its eclipse by the I-16. Speed, dive and climb were becoming the most important attributes of the fighter, and by the end of its career, the I-15's incredible manoeuvrability had become an irrelevance. But for a year or two, the chunky Polikarpov was demonstrably one of the two best fighters in the world, bar none

Fiat CR.32 Freccia (1933)
'Rosatelli's arrow'

TOP SCORER: Joaquín García-Morato y Castaño – 36 kills flying the CR.32, of a total 40

The CR.32 and its great rival, the Polikarpov I-15, were the main aggressors in the Spanish Civil War, the last major conflict in which biplane fighters could still be regarded as the finest combat aircraft in the world, Even then, their time at the top was relatively brief and they were supplanted by the Fiat G.50 and I-16 monoplanes, respectively. The Fiat was the pinnacle of a line of excellent fighters designed by Celestino Rosatelli for a flamboyantly confident Italian air force. Oddly, both the Fiat and its Polikarpov nemesis fought as part of the Chinese air force against Japan. In China, the CR.32 was found to be superior to both the Curtiss Hawk and Boeing P-26 'Peashooter', but operational use in the country was relatively limited due to the scarcity of its fuel – its 600 hp Fiat A30 engine required a heady cocktail of alcohol, benzole and petrol to work properly – and other, less picky fighters came to the fore.

Meanwhile, over Spain, the Fiat flew with the Fascists against the Communist upstart I-15, its one major advantage over the Soviet aircraft being its heavier armament. The bitter tragedy of the Spanish Civil War would prove to be the CR.32's finest hour: it was able to destroy the Tupolev SB-2 bombers that were thought uninterceptable due to their speed. Extraordinarily, both sides possessed monoplane bombers that were faster than their respective biplane fighters, a situation that hastened the latter's replacement by speedier monoplanes.

'Compared to the decidedly elegant CR.32, the I-15 looks like it has been driven into a wall'

'The Kawasaki proved master of all it encountered, until it ran up against the latest Soviet types during the so-called Nomonhan Incident'

Kawasaki Ki-10 (1935)
'Imperial acrobat'
Top scorer: Kosuke Kawahara – all 8 kills while flying the Ki-10

Virtually unknown today, the Kawasaki Ki-10 – code-named 'Perry' by the Allies during the Second World War – established an air superiority over China that would last for years. It gave the Japanese their first air aces and firmly founded the near-pathological Japanese obsession with manoeuvrability above all other fighter attributes, which would ultimately lead to the exceptional Mitsubishi A6M Zero. It is also the only aircraft on this list from whose manufacturer you can today buy a brand-new motorcycle. Interestingly, the Ki-10 was judged to be superior to a monoplane competing for the same order – the Nakajima Ki-11. This would be the last time a fighter biplane would be selected in preference to a rival monoplane anywhere in the world.

In service, the Kawasaki proved master of all it encountered, until it ran up against the latest Soviet types during the so-called Nomonhan Incident, a decisive victory for the Soviets during the undeclared Soviet–Japanese border conflicts of 1939. By then, even Japan had accepted the inevitable and the Ki-10 was replaced by a developed version of the monoplane it had initially beaten, in the form of the dainty Nakajima Ki-27. Even then, the Ki-10 forged a secondary career playing Chinese aircraft in the excellent aerial sequences in the 1940 flick *The Burning Sky*.

Gloster Gladiator (1934)
'Gladiators ready!'
Top scorer: Marmaduke Pattle – 15 kills flying the Gladiator, out of a total 40-plus

Everyone already knows everything about the Gloster Gladiator, don't they? So, which nationality was the first pilot to score a kill in one? That's right: American. John 'Buffalo' Wong flying in the Chinese Air Force in 1938 shot down a Mitsubishi A5M in a Gladiator long before 'The Flying Tigers' had even been thought of. This kind of cosmopolitanism, its having been flown in all sorts of places by all sorts of unlikely operators, is typical of Britain's last fighting biplane; the Gladiator was little more than a convenient stop-gap to keep up the numbers until the Hurricane and Spitfire came on stream in sufficient quantity, so it was released for export at a fairly early date.

The Gladiator pops up in an unusual number of imbalanced

conflicts far from its home where it was forced to operate in the face of numeric and technological superiority – invariably heroically and to great propaganda value – thus pithily illustrating the general experience of the biplane fighter in the Second World War. The Gladiator stoically defied the odds, flying with the Chinese against the Japanese, the Finns against the Soviets, the Belgians against the Luftwaffe and, most famously, with the RAF against the Italians over Malta. More prosaically, when operated in numbers against a similarly equipped enemy it performed excellently, and a situation developed over Africa similar to the CR.32–I-15 one in Spain: it clashed regularly with the Fiat CR.42, which, though slightly faster, did not handle as well as the Gloster. Despite being the RAF's last biplane fighter, it was also that service's first fighter to sport an enclosed cockpit. There are not many aircraft that have been simultaneously in the vanguard of development while also totally obsolete.

Polikarpov I-153 (1937)
'Jonathan Livingston Chaika'

Top scorer: Aleksandr Avdeyev – 12 kills in the I-153, out of a total 13

Generally, the Polikarpov I-153 did not fare well in combat, but its inclusion here is justified not only because it took biplane performance to the furthest limits (at 280 mph, or 450 km/h, it was the fastest biplane to see service) but also because, obsolescence at service entry notwithstanding, on one occasion it replaced the seemingly superior I-16 monoplane on operations. The I-153 was so good that even in a monoplane-infested world, it persisted in frontline service until 1945.

Given that the Soviet Union was the first nation to introduce a modern monoplane fighter into its inventory, as early as 1934, it may seem odd that they should persist with the biplane, but Soviet tactical thinking foresaw a tidy combat situation wherein monoplane fighters would break up a force of incoming bombers, leaving the I-153s to deal with them, and any pesky escorts, individually. In reality this didn't work. Nonetheless, 3,437 I-153s were built and were heavily used, mostly in a desperate rearguard action against the invading Germans in 1941. What's more, back in 1939, the brand-new I-153 was rushed to the Mongolian front to replace the I-16 monoplane. The Japanese were fielding the Nakajima Ki-27, which possessed the sparkling agility to outmanoeuvre the faster I-16. The I-153 offered near parity in performance and manoeuvrability terms, and combat performance against the Ki-27 quickly improved.

Interestingly, the I-153 is also one of a handful of fighters to fight itself. In March 1943, two Soviet I-153s clashed with three Finnish I-153s; with one aircraft forced to land after suffering damage in the ensuing mêlée. The Finns, never ones to ignore a decent aircraft, operated about a dozen captured I-153s against their former owners until February 1945, by which time the eight surviving aircraft were the last biplane fighters flying on operations anywhere in the world.

K 8032

K8032

TOP 10
FIGHTERS OF THE SECOND WORLD WAR

Fighter development in the Second World War was a Darwinian bloodbath that would have had Richard Dawkins slavering with excitement. A brutal survival of the fittest ensured a rapid evolution of these characterful machines. With up to twenty-five times more horsepower, more than eight times the firepower and quadruple the speed of the fighters of the First World War, the new generation proved brutal engineering masterpieces. Never before had fighter aircraft been as vitally important for the fate of so many nations, and nor have they since. Here's our selection of the ten best.

10. Grumman F6F Hellcat (1942)
'Steamboat Fatty'

The Hellcat was ordered as a lower-risk back-up in the event of any major problems with the Vought F4U Corsair, which was very prudent as the Corsair programme very quickly ran into major problems, and the sturdy F6F found itself the premier carrier fighter in the world's mightiest carrier fleet. The Hellcat was big, heavy and extremely powerful, the very antithesis of its major opponent, the A6M Zero.

To fight the Zero, pilots of earlier Allied naval fighters had had to devise inventive tactics to deal with the superior Japanese aircraft. But with the Hellcat, the US Navy had a fighter that was slightly faster, better armed and just manoeuvrable enough to deal with the Japanese plane. It was also extremely strong and easy to fly, factors that saved many a pilot who would have been doomed in any other aircraft. The Japanese advance had been checked by the Hellcat's predecessor, the F4F Wildcat, but it was the chunky F6F that allowed the US Navy to win the war in the Pacific – it was exactly the right aircraft at exactly the right time. It was replaced only right at the end of the war, in part by its old nemesis, the Corsair.

ABOVE: Space is at a premium on a carrier, and the Hellcat was an unusually massive example of its type, boasting a larger wing area than any other WW2 single-engine fighter.

BELOW: Demonstrating its ferocious firepower, this P-39 lets rip with the awe-inspiring combination of a nose-mounted 37-mm cannon and four .50-calibre heavy machine guns. At low level, the P-39 was a dangerous opponent to underestimate.

'The Hellcat was big, heavy and extremely powerful, the very antithesis of its major opponent, the A6M Zero'

9. Bell P-39 Airacobra (1938)
'Dear little cobra'

When the P-39 first flew, it had a turbosupercharger and was a fantastic performer at all altitudes. However, the US Army Air Corps decided that no fighter would ever be required to operate at high altitude so they removed the turbosupercharger and developed the P-39 to make it a low-altitude fighter par excellence. Then, when it was committed to combat, the same US Army Air Corps were scathing in their criticism of the P-39's altitude performance and called it 'especially disappointing'. A bit rich, you might think, seeing as they were the ones who had cacked it up in the first place (though, to be scrupulously fair, Bell had experienced difficulties with turbocharger development and were also keen to drop this feature). Thus, the unwanted Airacobra was sent by the thousand to the Soviet Union, where it found itself in a battlefield where virtually all combat was at low level and its capabilities could be properly appreciated. It was fast (a P-39 unexpectedly won the first post-war Thompson Trophy, beating P-51s and P-38s, types generally viewed as superior), it handled beautifully, it was tough, its tricycle undercarriage was perfect for rough field operations and its firepower was nothing short of spectacular. Of the six Soviet pilots to score more than 50 kills, four were flying the P-39 – its performance was certainly superior to the German aircraft it faced (and the Soviet aircraft it complemented). The Airacobra gained more air-to-air kills than any other US-built fighter over Europe and demonstrated the remarkable strategic wisdom of the 'planes now, money later' Lend-Lease programme. Given that the Eastern Front used up 80 per cent of the German war effort, the so-called 'Kobrastochka' ('little cobra') could reasonably be considered the most important American fighter of the war in Europe.

ABOVE: Messerschmitt combined all the most advanced technologies of 1934 to design a stressed-skin fighter with a retractable undercarriage and the smallest and lightest possible airframe around the most powerful available engine.

8. Messerschmitt Bf 109 (1935)
'Cheap and deadly'

The 109 was arguably the best fighter in the world from the point of its introduction until about 1942, despite being 'a pile of shit' from a construction point of view, according to an aircraft-restorer–engineer friend of mine. However, it was also very cheap and easy to manufacture and it was this that led it to become the most-produced fighter ever (or second-most produced, depending on your criteria – see No. 7 in this list). Even once it had passed its developmental zenith, it represented a potent foe and was never totally outclassed by its opponents. The 109 scored more air-to-air kills than any other aircraft before or since its service, and probably represents the best value for money of any fighter in history. Try saying that about the F-22.

SPEER'S POTEMKIN AIR FORCE

Albert Speer is credited with the miracle of German fighter production in 1944, when vast numbers of such aircraft were built. But when Allied investigators interrogated Speer and started adding up the numbers in his department's production figures, they discovered a remarkable secret: 8,000 German single-engine fighters in the ledgers didn't exist. Further investigation revealed that, although Speer had managed to dramatically increase the number of fighters produced, he had also cooked the books to gain favour with Hitler. Speer had done it by reallocating all aircraft that were being repaired or refitted to the 'new aircraft' ledger, thus giving a dramatically overinflated impression of his achievements. That wasn't all. German engine designers told Allied engineers that the impressive final boost-levels released by Daimler-Benz for the Bf 109, of +2.1 and even +2.3 atmospheres manifold pressure, were needed just to get the 109 to meet its basic service specifications. Speer's 'miracle' had created fighters of such incredibly poor build-quality that the Daimler-Benz engine designers revealed to the Allies that the fighters reaching the front line were on average an incredible 25 mph (40 km/h) – over 6 per cent – slower than their claimed performance. The Allies had faced hordes of 'ghost fighters': those that were not figments of Speer's ledgers were, in performance terms, mere shadows of their potential.

7. Yakovlev Yak-1 to -9 (1940)
'The triumph of socialist labour'

When the great Soviet ace Alexander Pokryshkin was being pressured for political reasons to convert his unit to a Soviet-built aircraft rather than the Airacobra he was then flying (to great effect), one of the proffered types was the Yak-3.

However, Pokryshkin detested the aircraft's designer, Alexander Yakovlev, and refused the offer of his latest fighter model. This was unfortunate, as by doing so Pokryshkin cheated himself out of flying the finest Soviet fighter of the war. The French Normandie-Niemen unit, which fought as part of the Red Army, had rather a different opinion. Allegedly, at the end of the war they were offered their choice of any Allied fighter aircraft and they selected the Yak-3. Marcel Albert, their top scorer, maintained that the Yak could outclimb a Spitfire and had a higher cruising speed.

The Yak-3 was one of a family of fighters that had begun with the Yak-1 and diversified into different, concurrently developed lines. Despite their different designations, there was less of a difference between the types than between an early- and late-model Messerschmitt 109, which adds weight to the argument that the Yak family as a whole can be considered to be the most-manufactured fighter of all time, with around 38,000 built in total. The Yak-3 was the lightest and smallest fighter to be used in numbers by any combatant during the war and this allowed it to give a remarkable performance on a relatively low-

powered engine. Despite its daintiness, the Yak-1 was, in terms of effectiveness, on a par with contemporary Bf 109 and Focke-Wulf Fw 190 models, and by the war's end it was comfortably superior to both. Unburdened by the extraneous equipment deemed necessary in the West, the Yak was a very pure sort of fighting machine and probably the most pleasing aircraft from a war pilot's perspective. And which other first-line 1940s fighter has had production restart for the civilian market in the 1990s or been modified into a basic trainer?

'The Yak-3 was one of a family of fighters that had begun with the Yak-1 and diversified'

TOP: Tiny and exceptionally agile, the Yak-3 punched well above its weight. Along with the Yak-7 and Yak-9, it re-entered production between 1991 and 2002. This picture shows an example of one of the recently-produced aircraft, a Yak-3M. In 2011, a Yak-3UPW set the official international speed record for a piston-engined aircraft in the under-3,000 kg (6,600 lb) category, at 407 mph (655 km/h).

ABOVE: The second of the Yak variants to see service, the Yak-7 started life as a trainer version of the Yak-1. Weirdly, engineer K. V. Sinelshchikov at state factory Zavod 301 altered a standard Yak-7UTI trainer back into a fighter. Flight testing revealed the new version superior to the Yak-1 while retaining the useful ability to carry a passenger behind the cockpit. In early-production Yak-7s, the rear cockpit was faired over, endowing it with a distinctive humpback profile.

'It was so good that a team of German experts (including the 104-kill Luftwaffe ace Adolf Galland) came to the conclusion that it was the best fighter in the Axis, possibly in the world, and should be produced in vast numbers immediately'

ABOVE: The excellent qualities of the Fiat fighter would see it outlive the Fascist state that created it and return to production after the end of the war, and G.55s would see action once again in Egyptian hands. Eventually Fiat re-engined the G.55 airframe with the Rolls-Royce Merlin to create the G.59, which served until the mid-1960s.

OPPOSITE: When eighty-seven-victory ace Hiroyoshi Nishizawa was photographed flying this A3M3 Model 22 in May 1943, the Allies had largely taken the measure of the superb Zero. The failure to develop a replacement for or significantly improve Mitsubishi's most famous product cost the Imperial Navy dear.

6. Fiat G.55 Centauro (1942)
'I could have been a contender!'

Just before everything went completely awry for the Italians, they managed to obtain a supply of the latest Daimler-Benz 605 engines from Germany and built three superb fighter types. All three saw service but the best was the Fiat G.55. Indeed, it was so good that a team of German experts (including the 104-kill Luftwaffe ace Adolf Galland) came to the conclusion that it was the best fighter in the Axis, possibly in the world, and should be produced in vast numbers immediately, for German use. Kurt Tank, designer of the Fw 190, had nothing but praise for the G.55 and went to Turin to look at its potential for mass production. Sadly for the Axis, cold, hard economic logic came into play and when it was pointed out that the Fiat, though admittedly outstanding, took 15,000 man-hours to build against the 5,000 of the still formidable Bf 109, mass production plans were quietly abandoned. Thus, fewer than 300 of the Axis's best fighter were built and saw service only in a backwater of the conflict for a Nazi client-state, whereas some 35,000 109s swarmed all over Europe. However, in contrast to so many potentially terrific might-have-beens of the war, the Fiat did at least see production and serve in combat, where its brilliance was demonstrated rather than merely conjectured.

5. Mitsubishi A6M Reisen Zero (1939)
'Shatterer of self-delusion'

Quick quiz question: what links the M16 assault rifle and the Mitsubishi Zero? That's right: 7075 aluminium alloy. It's used for parts of the M16, and for most of the structure of the Zero. First produced in 1936 by Sumitomo Metal Mining of Japan and excitingly named 'extra super duralumin' at the time, 7075 is an alloy of aluminium and zinc and is significantly lighter and stronger than the other aluminium alloys that had been produced before this date. The fact that the A6M's designer, Jiro Horikoshi, had to resort to new technology at a metallurgical level demonstrates not only how challenging the Japanese Navy's specification for their new fighter was (Nakajima Aircraft Company didn't even enter a tender, as they deemed it impossible) but also how cutting-edge the Zero was for its time. Particularly demanding were the requirements relating to range, maximum speed and agility.

At its debut, the A6M was the world's best carrier fighter. This was totally ignored by the Allies, despite the aircraft being quite openly used over China, which suggests that the West was all too willing to believe its propaganda about the weakness of Japanese military capabilities. Today that propaganda seems at best laughably naive, and at worst founded on an illusion fostered by dogged racism. Whatever the truth, the Zero changed all that, and with such

total dominance that for the first couple of years of the Pacific War opponents came to believe the Japanese were invincible. By the time American fighter design had caught up a bit, grubbily specious claims were invented to 'explain' the Zero's remarkable design and performance. For example, the pee-collecting millionaire Howard Hughes claimed after the war that Mitsubishi had copied his H-1 racer, and Eugene E. Wilson, then president of the aircraft manufacturer Chance Vought, claimed they had copied the Vought company's own V-143 fighter (which was in fact rather mediocre). It is bizarre that industrialists should resort to lying about an enemy aircraft ('We designed it, really!') in a war already won but it also demonstrates the Zero's absolute superiority and just how infuriating this was for the grandees of the US military-industrial complex. Praise indeed.

4. Focke-Wulf Fw 190 (1939)
'The Butcher Bird'

Quite apart from being an excellent aircraft with several radical features, the Fw 190 heralded a revolution in what today would be called 'ergonomics' (or 'human factors') and that in 1941 was often dangerously overlooked.

Today the concept of HOTAS (hands on throttle and stick) is well known and is generally considered to have been pioneered by the F-16. However, the 190 sported a system that delivered a kind of proto-HOTAS concept some thirty years before the Viper. Known as the Kommandogerät ('command device'), it was a remarkable mechanism that automatically controlled fuel flow, fuel mixture, propeller pitch and ignition timing. It also activated the supercharger at the correct altitude – all the pilot had to do was move the throttle lever. His other hand was on the control column, where all the armament switches were situated, allowing his full awareness to be directed to combat. This awareness was further enhanced by the aircraft's bubble hood, which offered a view that one contemporary RAF report described as 'the best that has yet been seen'. When one considers that, on its debut, the Fw 190 was superior to its closest rival, the Spitfire V, in every performance parameter (except turn rate), yet also gave its pilots a tactical edge thanks to reduced workload, it's no wonder that its very existence sent British designers into a frenzy of activity to try and regain ascendancy. The Spitfire and other fighters later achieved parity but the Fw 190 remained a dangerous opponent and, like the F-16, saw its main role shift to place greater emphasis on its fighter-bomber role. An aircraft that defined the state of the art, the Fw 190 could be considered the first truly modern fighter.

ABOVE: At the time of its service entry, the Fw 190 was certainly the best operational fighter aircraft in the world. Its deadly superiority was a nasty surprise to the RAF fighter pilots flying against it.

SEXING DOWN 'THE BUTCHER BIRD'

The Fw 190 was judged to be such a serious threat when it arrived that Chief of the Air Staff Sir Charles Portal wrote to Churchill on the matter. The archives show that he rewrote the letter twice in March 1942 before sending it, each successive letter putting the threat across in increasingly euphemistic language. The first read: 'adverse casualty rate', the second version scored out this line and replaced it with 'unfavourable balance' and finally 'unfavourable factor'.

3. North American P-51 Mustang (1940)
'Little friend'

Everyone's been going on for years and years about how incredibly fantastic the Mustang was, but this tends to distract from what actually made it such a remarkable aircraft. It's worth remembering that it wasn't supposed to have existed at all, and came about solely because North American Aviation didn't particularly want to build Curtiss-designed P-40s for the British. Even then, the Mustang would have been a competent but hardly spectacular improvement on the Curtiss fighter had not some bright spark suggested fitting it with a Merlin engine (curiously, this step was taken independently yet almost simultaneously by North American in the US and Rolls-Royce in the UK). Nonetheless, many pilots were initially less than impressed, citing the finer flying characteristics of the Spitfire and the better build quality of the Republic P-47. But the Mustang was at least as good a fighter as either and could fly to Berlin and back. 'When I saw those Mustangs over Berlin, I knew that the war was lost,' said Hermann Goering, and he was right. Whether or not it was the best fighter of the war, the Mustang remains the standard against which all other hopefuls are judged.

BELOW: The superior aerodynamics – notably the laminar-flow wing – of the P-51D made it faster on the same engine than a Spitfire IX. A far larger internal fuel capacity blessed it with much longer range, too.

2. Supermarine Spitfire (1936)
'The Untaming of the Shrew'

If you ask people today what a 'spitfire' is, virtually everyone – in the UK, at least – will answer 'an aeroplane'. It is extremely unlikely that anyone would say 'someone with a fierce temper', despite their slightly tedious accuracy if they did. This is the enduring legacy of the Spitfire: it has become the definition of the word originally used to name it; its success has changed the language. There is only one other aircraft I can think of that has done this, and that is Concorde (which uses the French spelling of the word). The Spitfire changed history, not just in the conventional sense of influencing the course of the war, but also in the way that the legend surrounding it has given it the starring role in the action, when in reality it was more of a supporting player.

But who cares? The story is better this way, the dumpy Hurricane relegated to being championed by boring aviation geeks while the eternally handsome Spitfire swans about oozing the sex appeal that any self-respecting fighter aircraft should have. Just look at it. Legend has it that the designer of the Spitfire, R. J. Mitchell, said of the wing shape: 'I don't give a bugger whether it's elliptical or not, so long as it covers the guns,' despite the fact that on every other fighter the guns were covered effectively enough without aesthetically lovely cladding that was also difficult to manufacture, so I suspect he was lying. Add to this the fact that the Spitfire was the only British fighter to remain in production for the duration of the war and to have comfortably remained one of the top five fighters worldwide throughout that time.

The Spitfire also achieved the highest speed ever attained by a piston-driven aircraft, which is pretty exciting. But the Spitfire doesn't need facts – its claim to being a contender for the best fighter of the war has nothing to do with tawdry reality and everything to do with the myth. The Spitfire effectively engendered its own legendary status.

BELOW: The Spitfire remained in service, with a variety of air arms, long after the Second World War. Israel bought fifty Mk.IXs from Czechoslovakia and the Spitfire's final air-to-air 'kills' were scored on 7 January 1949 when Israeli Mk.IXs shot down three British Spitfires and a Hawker Tempest. The first two Spitfires were shot down because they were erroneously thought to be Egyptian. The other was hit along with the Tempest in an engagement between RAF Spitfires and Tempests and Israeli Air Force Spitfires. The Israelis assumed hostile intent from the British, having mistaken the Tempest drop tanks for bombs, and destroyed two British planes for no loss. Due to maintenance failures, the RAF Spitfires couldn't fire their guns and the Tempests couldn't release their drop tanks.

'The Spitfire changed history, not just in the conventional sense of influencing the course of the war, but also in the way that the legend surrounding it has given it the starring role in the action, when in reality it was more of a supporting player'

WHY GERMANY MISSED OUT ON THE TWO-STAGE SUPERCHARGER

The Mk.IX Spitfire, which had a two-stage supercharged engine, is lauded for having rescued the RAF from the clutches of the Focke-Wulf Fw 190. Given the new Spitfire's effectiveness, did the Focke-Wulf rapidly counter it with a similarly equipped fighter? No. Germany fitted two-stage superchargers to only a couple of fighter types, the Fw 190 D-11 and Ta 152, powered by the Junkers Jumo 213E and F, and both entered service far later, close to the end of the war. The Jumo 213E had a system like the Merlin, but appeared only in tiny numbers a few months before the collapse of Germany.

Germany's failure to develop a two-stage supercharged fighter is widely regarded as one of their great technical failures in high-altitude flight. However, this was as much an organisational failure as anything else, as Daimler-Benz had been running two-stage supercharged DB 601 engines since around 1936 in their test cells in Stuttgart-Untertürkheim. Focke-Wulf master-designer Kurt Tank told incredulous American engineers that the notion of a two-stage supercharged fighter had been quashed by German military bureaucrats before it could take full form. The Luftwaffe was regarded at the time as purely a tactical force, designed to support ground operations. They saw the ability of fighters to attain very high altitudes as anathema to this strategic concept, and cancelled all such projects. They in fact had their own equivalent to the Spitfire Mk.IX's engine, but wasted the opportunity through sheer ineptitude.

TOP: Considering it was designed specifically as a short-range interceptor, the versatility of the Spitfire was remarkable. This Irish example started life as one of the Royal Navy's Seafires before being 'de-navalised' to serve as a general-purpose fighter. Irish Seafires remained in service until 1955.

WHY, REALLY, DID THE GERMANS PUT A GERMAN ENGINE ON A SPITFIRE?

The now well-known 'MesserSpit', or 'German Spitfire', was a captured Spitfire Vb fitted with a Daimler-Benz DB 605 engine. The change of engine was largely to settle a feud between the head of Daimler-Benz, Fritz Nallinger, and Professor Willy Messerschmitt.

In 1943, Messerschmitt, furious with being blamed for the poor performance of German fighters compared with the latest Allied types, told Erhard Milch (Germany's head of aircraft production and supply) that this was no surprise to him because he had been forced to fit water radiators twice the size of those the Spitfire used per horsepower delivered. Milch, astonished, turned to the head of German engine development, Wolfram Eisenlohr, and asked him: 'How have our designers not noticed this?' Germany had failed to develop high-pressure, high-temperature water-cooling. This meant that their radiators had to be significantly larger in section than those the Allied fighters used, adding to their drag values. They estimated this was costing German fighters at least 15mph at top speed, enough to turn a performance edge into mere equality.

1. Messerschmitt Me 262 (1942, with jet engines)
'Sturmvogel warning!'

There was quite a lot wrong with the Me 262 when it was committed to action but most of this was due to the exigencies of the time; aside from this, it represented an astounding technological advance when it was unleashed on an unsuspecting world in the spring of 1944. The obvious advantage of its new powerplant, the jet engine, was velocity. Once airborne, no other aircraft could catch the speedy Messerschmitt, not even the Allies' jet the Gloster Meteor, whose performance was decidedly pedestrian by comparison. But it wasn't just the 262's jet engines that made the fighter so formidable; its firepower, optimised for bomber destruction, was particularly heavy, consisting of four 30-mm cannon firing explosive rounds at an extremely high rate. The 262 was also in some senses a remarkably practical aircraft for the not particularly advantageous situation into which it was introduced. It could be fuelled by a much lower quality of fuel than its piston-engined brethren so there was more chance of being able to operate it in oil-starved Germany. Furthermore, a surprisingly large amount of the airframe was made of wood rather than ever-more-scarce aluminium and steel. Scarcity of steel was the main cause of its biggest problem: the engines

were notoriously short-lived because steel of sufficiently high quality was no longer available for the turbine blades. It's also worth remembering that these engines, as well as the airframe, were built by slaves, so it's hardly surprising that the build quality wasn't that great – in fact, it's remarkable that the aircraft worked at all. But even with these niggles, the 262 reigns supreme as an incredible technological last gasp at the end of a war already lost. Its very existence heralded a new age in fighter design. It was as if it had popped up from the future to astound and astonish. There is even some suggestion that it may have broken the sound barrier.* The Messerschmitt 262 was in a class of its own. Not bad for an aircraft that was supposed to be a bomber.

---

* There are unconfirmed stories of Luftwaffe pilots reaching the Mach in the German Me 262, the first jet fighter. Some claim German test pilots went supersonic in prototype rocket-propelled fighters. The stories about the Me 262 breaking the sound barrier, even in a dive, are almost certainly mistaken, however. The Me 262's fuselage design and fat wings would be considered incompatible with supersonic flight today. Cockpit airspeed indications sometimes jump around in transonic flight as shock waves form near the pitot tube, and this may have led some pilots to believe they'd exceeded the Mach.

ABOVE: As a swept-wing jet fighter, the Me 262 was a vision of the future. Astonishingly fast and extremely well armed, once airborne it offered a nightmarish opposition for Allied pilots.

THE WORST POSSIBLE AIRFORCE

The illustrator Edward Ward approached Hush-Kit complaining that life seemed pointless. As he sat in the park in the rain, bemoaning the downfall of civilisation, I decided he needed something to distract him. With this in mind, I gave Ed the enviable task of equipping a notional 1940s air force – with one proviso: he could only pick from the worst aeroplanes flying at the time. Over to Ed. . .

Picture the scene: It is the 1940s. The Republic of Hushkonia has been taken over by a benevolent dictatorship of disgruntled aviation enthusiasts. Somewhat ironically, the Hushkonian Air Force (and the national airline, Air Hush) remains under the control of officers loyal to the old regime. Furious with their new ineffectual overlords, yet too timid to stage a coup, they decide instead to make the Hushkonian armed forces as bad as possible in an attempt to encourage takeover by a foreign power and restoration of the old order. To add at least some credibility to their actions they decide to select aircraft that actually saw service in their specified roles with other nations. No crazy prototypes or mad schemes here, only tried and tested flops.

Fighter: Messerschmitt 163

'Hello, Messerschmitt? Luftwaffe here, we'd like a new fighter please.'

'Righto. What kind of thing would you like?'

'Well, we were hoping for something that looks really cool and everything, and is really, really fast – mmm, in fact, would it be too much to ask for it to be fast enough for the closing speed between it and any potential target to be so high that it would be more or less impossible to aim and fire the guns at anything with any realistic chance of success? And could it have a cannon with a really low muzzle-velocity, to properly compound that problem? Also, we were wondering if it might have no range at all, and it'd be good if we could have it land on a ski or something, preferably as a glider, and we want it to blow up all the time for no apparent reason, so it'd be best if it was full of crazy volatile fuels. Oh, and if possible, we'd like the fuel to dissolve the pilot. Talking of the pilot, we thought it might be nice to design in a terrifying aerodynamic flaw that will definitely kill him, like maybe an unrecoverable dive if he lets the aircraft exceed Mach 0.84.'

'Would you like it to be pressurised?'

'No.'

'Ejection seat?'

'No.'

'Anything else?'

'Did we say we wanted it to look really cool?'

On the basis of this conversation, which actually, really, actually happened, the fighter arm of Hushkonia was equipped with its premier air-superiority asset.

OPPOSITE: The Fairey Battle was a great way to waste a Rolls-Royce Merlin that you could have put in a better aircraft instead. Note the sporting action of the pilot in wearing a dazzling white flying suit to make him a more obvious target.

ABOVE: Whereas most fighters of the time had a climb rate of around 3,000 feet per minute, the rocket-powered Me 163 clawed upwards at an astonishing 16,000 feet per minute. Pilots were ill-equipped to understand what was going on at such extreme climb rates and speeds.

'No aircraft has ever flown with such a formidable defensive armament. Unfortunately, this made the aircraft so draggy and heavy that it couldn't keep up with the bombers it was supposed to be protecting'

Long-range escort fighter: Boeing YB-40 Flying Fortress

This fighter is, as you have no doubt spotted, a B-17. Imagine mixing it with the 109s in this. In 1942, the US Eighth Air Force thought they might create an effective escort by slinging a massive number of guns onto a bomb-free Flying Fortress. No aircraft has ever flown with such a formidable defensive armament. Unfortunately, this made the aircraft so draggy and heavy that it couldn't keep up with the bombers it was supposed to be protecting. In a totally irrelevant but oddly satisfying aside, the YB-40 is the only aircraft on this list to feature in an Oscar-winning film – two of them appear in the scrapyard scene at RFC Ontario towards the end of William Wyler's *The Best Years of Our Lives*, which won nine Academy awards in 1947. The YB-40's film career was notably more successful than its operational one but did not save it from the scrapman's torch.

ABOVE: The YB-40 could be equipped with up to thirty defensive guns, though it normally carried between fourteen and sixteen. Armament was mostly the .50-calibre M2 Browning machine guns in various configurations, though 40-mm cannon were also tested.

Strategic bomber: Heinkel He 177 Greif ('Griffon')

When it comes to long-range strategic bombers, there's really only one choice: the *Luftwaffenfeuerzeug*, Heinkel's flaming coffin, the homophonically accurately named Greif. It's worth pointing out that when it worked properly, the He 177 was a stupendous performer, powerful and fast. The trouble was that it seldom did work properly. Furthermore, when things started to go wrong in a Greif, they tended to go wrong quickly, catastrophically and inflammably. The statistics are enlightening: for example, of thirteen missions flown on flak-suppression duties at Stalingrad in 1942–43, seven 177s were lost to fire, none of which were attributable to enemy action.

The problem all stemmed from the He 177's powerplant, consisting of a pair of Daimler-Benz V12 engines mounted on a common crankshaft in each wing, which had an incredibly tight fit into their cowlings. Both engines shared a common central exhaust manifold serving a total of twelve cylinders – the two inner cylinder banks of the component engines. This central exhaust system would often become extremely hot, causing the oil and grease that routinely accumulated in the bottom of each engine cowling to catch fire. This problem was compounded by the fact that the fuel-injection pump on each engine tended to lag in response to the pilot throttling back in such situations, delivering more fuel than was required and thus fuelling the fire. In addition, the fuel-injection-pump connections often leaked. What's more, to reduce the aircraft's weight, no firewall was provided, and the back of each engine was fitted extremely close to the main spar, with two-thirds of each engine being placed behind the wing's leading edge. This meant that fuel and oil fluid lines and electrical harnesses were crammed in together with insufficient space, and the engines were often covered with fuel and oil from leaking fuel lines and connections.

And there's more: at high altitude, the poorly designed lubrication pump led to the oil foaming, reducing its lubricating qualities. Insufficient lubrication ultimately resulted in the failure of connecting-rod bearings (a problem that also befell the Avro Manchester, but that aircraft was quickly altered to become the superlative Lancaster), which led to the con-rods sometimes bursting through the crankcases and puncturing the oil tanks, the contents of which would then empty onto the white-hot central exhaust manifold. The tightly packed nacelles in which the engines were installed on the He 177, with many of the engine's components buried within the wing, led to very poor ventilation as well as poor maintenance access. Essentially, the He 177 was

a fire waiting to happen. While the repeated fires were by far the most serious issue affecting the Greif, the big Heinkel also had to contend with an overly heavy undercarriage, a dangerous swing on take-off due to the massive torque of the enormous propellers, an inadequate defensive armament and some unpleasant handling characteristics. On top of which, famed test pilot Eric Brown suggested the elevator control was 'dangerously light' and there lingering concerns about its structural strength, Brown noting that 'it really was nail-biting to have to treat a giant like this immense Heinkel bomber as if it were made of glass'.

The French finished a version of the He 177, with four separate engines, after the war, and it served reliably for years on test programmes, proving that if Heinkel hadn't inexplicably persisted with the coupled-engine concept, they could have had an effective, dependable strategic bomber from 1942. An amazing 1,169 of these terrible bombers were built. However, slightly sadly, none survive (except, of course, for the thousands being built for Hushkonia).

'The French finished a version of the He 177, with four separate engines, after the war, and it served reliably for years on test programmes'

BELOW: The derided He 177 was one of the aircraft that most influenced the course of the Second World War. Unfortunately for Germany, it was influential in exactly the opposite direction to that which they intended. Named after the mythical griffon, the mighty aircraft promised much and delivered considerably less. Unlike dragons, griffons do not breathe fire, but fire was to prove particularly pertinent to the He 177 programme.

Light bomber: Fairey Battle

Every fighting power of the Second World War pulled out all the stops to produce dreadful light and medium bombers apparently designed solely for killing aircrew, but the Battle lowered the bar of uselessness to an unassailable depth. Despite being the first RAF aircraft to shoot down an enemy aircraft in the Second World War, and the first aircraft to be fitted with the superlative Merlin engine, the Battle was woeful. It was a kind of anti-Mosquito, being too slow to evade enemy fighters yet too poorly armed to defend itself, too small to carry a decent bomb-load yet too large for a single-engined aircraft, and lumbered with an extra crewman to no real purpose. The Battle was unable to survive against any modern fighter aircraft and loss rates during 1940 regularly exceeded 50 per cent and reached 100 per cent on at least two occasions. It does not require a degree in mathematics to realise that losses at these levels are untenable.

Its shortcomings had been recognised before the war, but the Battle had one overriding trump card: it was cheap. In late-thirties Britain, it was decided that to have lots of crappy bombers was better than having none at all, especially when announcing production totals to a hostile parliament and press. It is not a coincidence that more Battles were built than any other aircraft on this list.

To be fair to the Battle, its contemporaries the Blenheim, Hampden and Wellington were also cut to pieces by day – suffering far fewer losses by night – but even if it managed to survive, the Battle's weedy bomb-load made it the most ineffectual of the lot. In the end it found its niche as a training aircraft, being of a useful size, reliable and free of vices, and it remained in use until 1949. In operations, however, it was a deathtrap put into service in large numbers for cynical reasons of economy and political disinformation. As such, it is the ideal light bomber for the Hushkonian air force.

ABOVE: In its defence, the Battle was undoubtedly a much better aircraft than the elegant biplane Hawker Hart it replaced, being able to carry a larger bomb-load further and faster.

BELOW: Pictured in flight (a rare occurrence) the sleek Ba. 88 certainly looks the part, but this handsome exterior belies an aircraft that was effectively useless. Surprisingly, the Lince had set several speed records in the late thirties; it was the addition of military equipment that pushed its weight up beyond the level its engines could cope with.

Ground attack: Breda Ba. 88 Lince ('Lynx')

Do you like aircraft that can go round corners? Breda thought that was overrated.

Proof that the adage 'If it looks right, it'll fly right' is a load of old cobblers, the Lince looked fast and purposeful yet it was so overweight, draggy and underpowered that it sometimes refused to fly at all. On the first day of the Italian offensive against British forces in Egypt, for example, three Bredas were committed from Sicily: one tried unsuccessfully to take off and another was found to be unable to turn and was therefore compelled to fly straight and level until it arrived at Sidi Rezegh airfield in Libya (which, fairly evidently, isn't Egypt). Later, once sand filters were fitted to the engines, the Lince could not exceed 155 mph (249 km/h) and there were occasions when entire units failed to take off. In an attempt to make the benighted craft viable, various items of equipment were left behind, including the rear machine gun, one of the crew (leaving the pilot all on his own), and half the fuel and bomb-load, but this never worked and the Lince was adapted to a role it fulfilled admirably – being parked on airfields to draw enemy fire. A noble task.

'In an attempt to make the benighted craft viable, various items of equipment were left behind, including the rear machine gun'

Reconnaissance: Curtiss SO3C Seamew

Proof that the adage 'If it looks right, it'll fly right' is totally accurate, the Seamew looked awkward and just, somehow, wrong. From the unlovely lines of its engine cowling to its horrible rectangular winglets and worryingly truncated rear fuselage, the Seamew inspired a total lack of confidence – with good reason, as it turned out, for the poor little Curtiss was a dreadful aircraft. It didn't even win the competition that selected it for service – a rival design by Vought was judged superior but Vought were busy with the F4U Corsair and Curtiss had spare capacity, so into production it went and in no small terms, as 795 of these unpleasant little aircraft were released into the wild. If it had been merely slow and uninspiring, it could be written off as a humdrum mediocrity, but the Seamew was also dangerous to fly.

Even if the Ranger engine didn't pack up (which it often did – a bad start for a single-engined aircraft intended to operate mainly over the sea), the plane couldn't take off with more than 80 gallons of fuel on board, despite the fact that its main tank could hold 300 gallons. The Seamew had other tricks up its sleeve, too: according to the improbably named Lettice Curtis, 'it was possible to take off in an attitude from which it was both impossible to recover and in which there was no aileron control', which certainly sounds stimulating! Eventually the Seamew became one of that select band of aircraft that were replaced by the very aircraft they were supposed to succeed, the biplane Curtiss SOC having been restored to the catapults of several US Navy capital ships. In an admirable gesture of inclusiveness, Curtiss made the SO3C available with either wheels or floats, so its unpleasant characteristics could be experienced equally by those on land or at sea.

ABOVE: Aircraft boldly being flung off catapults always look good, even if they're as dreadful as the Seamew. This angle also serves to obscure the bizarrely abbreviated fuselage and general aesthetic awkwardness of the design.

> 'Another eccentric American scheme named Project Pigeon utilised a pigeon in a bomb, trained to peck at an image of its target on a small screen, which in turn operated controls to keep the bomb on track'

AIR-DROPPED WEAPONS

But what should the massed ranks of Hushkonia's Fairey Battles and Heinkel 177s carry into the fray to strike terror into the hearts of their enemies (that is if they don't catch fire first)? The main weapon system of Hushkonia features the most terrifying animals from history: the bat and the pigeon. While the bat has a certain spooky cachet as the alter ego of choice for most modern vampires, the US-developed bat bomb made no use of the psychological aspect in its intended destructive process. The bomb consisted of a canister containing over 1,000 hibernating Mexican free-tailed bats (*Tadarida brasiliensis*), each fitted with a small napalm-filled incendiary device with a timer. The canister was dropped from a standard bomber aircraft, parachuting to a soft landing, whereupon the bats awakened and flew to roost under the eaves or attics of buildings within a range of about 40 miles (64 km), where the timed bombs they were carrying would, it was hoped, ignite and cause widespread fires. The bat bomb was the invention of a dentist called Lytle S. Adams, who rather uncharitably described the bat as the 'lowest form of animal life' and asserted that, until that point, 'reasons for its creation have remained unexplained'.

In fact, the sole damage inflicted by the bat bomb occurred during tests of the device in May 1943, when Carlsbad Army Airfield was accidentally set on fire after the armed bats roosted under a fuel tank and ignited it. Nonetheless, the bat bomb was judged to be very effective and it was only the advent of the atom bomb that prevented the bat bomb's use in action. Thus, the disarmingly unethical delivery system requiring the destruction of several million bats, and the fact that it remains, to coin a phrase, 'batshit mental', necessarily warrants its inclusion in the arsenal of Hushkonia.

Meanwhile, another eccentric American scheme named Project Pigeon utilised a pigeon (*Columba livia domestica*) in a bomb, trained to peck at an image of its target on a small screen, which in turn operated controls to keep the bomb on track. Despite demonstrating greater success than early electronically-guided weapons, project developer Dr Burrhus F. Skinner noted that 'our problem was no one would take us seriously'. Well, Hushkonia takes this very seriously indeed, not least for the delicious irony of employing a dove of peace as a weapon of war, and has amassed several thousand avian 'volunteers' for immediate training.

ABOVE: An expensive way to prove the effectiveness of your new secret weapon: this is Carlsbad Army Airfield Auxillary Air Base in New Mexico after being set alight by armed bats following the accidental release of several of the unfortunate avian mammals.

Carrier fighter: Blackburn Roc

The wrong concept applied to the wrong airframe at the wrong time, the Roc was the answer to a question that should never been asked, namely: 'Where's the Navy's Boulton Paul Defiant?'

Boulton Paul had gone to great lengths to make their turret-armed fighter as fast and handy as possible. Despite carrying around a turret and a gunner, which added about a ton to the loaded weight of the aircraft, the performance wasn't much worse than a contemporary Hurricane and, although the concept was flawed, the aircraft was excellent. Imagine what they must have thought when the Navy asked them to mount the same turret in the less-than-stellar Blackburn Skua to produce the Roc, which was 85 mph (140 km/h) slower and infinitely less able to survive, let alone fight, in the skies over Europe. Exactly how an aircraft derived from a dive-bomber, barely able to reach 200 mph (320 km/h) and with no forward-firing armament was supposed to combat a Messerschmitt 109 was apparently not a major concern for the powers that be.

Luckily for all concerned (except the Luftwaffe), the Roc was little used, but amazingly it did score one kill, against a Ju 88 over Belgium, an aircraft nearly 100 mph (160 km/h) faster than the unlovely Roc. Despite this unlikely success, the Roc remains the worst operational carrier fighter ever to grace a flight deck, and as such is the shoo-in for the noble Hushkonian fleet.

'The Roc was the answer to a question that should never been asked, namely: "Where's the Navy's Boulton Paul Defiant?"'

ABOVE: While it was adapted for ground attacks with bomb racks under the wings, the Roc was a hopeless air-to-air fighter.

BELOW: The Devastator looked pretty hot stuff when it entered service in August 1937, but four years later was proven to be catastrophically vulnerable at the Battle of Midway, where a total of forty-one TBDs were committed, and only six returned.

Carrier torpedo bomber: Douglas TBD Devastator

The Devastator's chronic vulnerability has become infamous. It was required to fly straight and level at a stately 115 mph (185 km/h) to deliver its torpedo, a speed that meant it could be easily intercepted by an SE5a of 1917 vintage, which is somewhat unfortunate for an aircraft touted on its debut as the most advanced naval aircraft in the world. By contrast, the contemporary Japanese Nakajima B5N could launch its superlative Type 91 torpedo at over 200 mph (322 km/h). Furthermore, the poor old TBD had a woeful defensive armament and lacked manoeuvrability. Its problems didn't stop there, as its main armament, the Mark 13 torpedo, was a dreadful weapon plagued with reliability issues and frequently observed to score a hit but then fail to explode. As a weapons system, the TBD–Mk 13 torpedo combination was probably the least satisfactory of the entire air war. Instead of the torpedo, the TBD could also carry 1,200 lb (540 kg) of bombs, thus extending the scope of its inadequacy into two roles. At least it could go a bit faster and higher when dropping its bombs, as they did not have the same deployment speed limitations as the Mk 13 .

Had it never been required to enter combat, the TBD would have been nothing more than another forgettable mid-thirties design. Dick Best, who flew a Douglas SBD dive-bomber at the Battle of Midway, remembered the Devastator as a 'nice-flying airplane', but, like the Fairey Battle, it was committed to combat in a world that had overtaken it. Only 130 were ever built, a pathetic amount for a US aircraft of this vintage, and, coincidentally, only six fewer than the equally dismal Blackburn Roc (see above). A match made in mediocre-naval-aviation heaven.

TOP 10

INCREDIBLE SOVIET FIGHTER AIRCRAFT THAT NEVER ENTERED SERVICE

Faced with such a mouthwatering menu of Soviet fighter projects that never entered service, it was almost painful to have to select a mere ten. No promises, but when the Hush-Kit writers are next sufficiently sober, we may create a Part Two.

'It is often claimed that the design therefore influenced the later Chinese/Pakistani FC-1 fighter, but there is significant evidence to suggest this wasn't the case'

10. Mikoyan Project 33 (never flown)
'F-20ski'

In the 1980s, the Mikoyan design bureau tinkered with a simple, single-engine warplane similar in concept to the original version of Lockheed's F-16 lightweight fighter. Like the F-16A, the new Soviet plane would be simple, manoeuvrable and inexpensive to produce.

The Project 33 design (sometimes – and perhaps erroneously – referred to as the MiG-33 or MiG-35) featured a single Klimov RD-33/93 afterburning turbofan engine, two of which powered the larger and more complex MiG-29. According to a 1988 report in *Jane's Defence Weekly*, Project 33 was 'seen as a complementary combat aircraft to the powerful MiG-29'. Where the MiG-29 boasted some multirole and beyond-visual-range (BVR) capability, the Project 33 was a short-range, point-defence fighter. Here was a MiG-21 for the 1980s – an ideal fighter for friendly states on a budget.

Project 33 didn't get very far, however, as Soviet leadership apparently preferred to devote the USSR's resources to more sophisticated aircraft. Mikoyan reportedly sold the Project 33 design to China after it became clear there would be no Soviet market for the plane. It is often claimed that the design therefore influenced the later Chinese/Pakistani FC-1 fighter, but there is significant evidence to suggest this wasn't the case. The design of the FC-1 (also known as the JF-17) was already quite far advanced at that point: originally conceived by the Chengdu Aerospace Corporation (CAC), it had then been developed through Sino–US cooperation with Grumman under the Super 7 name, and finally by CAC with Pakistani participation. When the Soviets stepped in, it was likely too late to achieve any substantial redesigns by incorporating elements from Project 33. To add to this, there is the possibility that the FC-1's design was instead more indebted to the Romanian IAR-95 fighter project, in which China was an initial player. Comparing a very early IAR-95 wind-tunnel model with an early FC-1 wind-tunnel model in Pakistan, the resemblance is much stronger than that between the FC-1 and any Project 33 model.

ABOVE: Had the aircraft entered service, it probably would have served with the Warsaw Pact air forces, in this case Bulgaria. Here we see an example as it could have appeared in the 1980s, armed with four unguided rocket pods.

9. Nikitin-Shevchenko IS-4 (1941)
'Конёк-Горбуно'

Picture the scene: it's the late thirties, you are aircraft designer Vasili Nikitin and you are puzzling out the future of the fighter aircraft while living in the terrifying day-to-day world of Stalin's Soviet Union. Yakovlev came up with a nice little fighter and was given a car; Polikarpov was a bit too cocky and was thrown in jail. And right now everything is awkward: the speed of the monoplane seems to be pointing the way to the future yet the biplane still has superior manoeuvrability, short-field performance and climb-rate. What the hell are you supposed to do? Suddenly, up pops apparently-crazy test pilot Vladimir Shevchenko, who explains over a couple of cups of *kvass* how you could achieve the advantages of the biplane and monoplane in the same airframe, with a hare-brained scheme he dubs the 'folding fighter'. Against all better judgement, the entire lower biplane wing hinges and retracts into the fuselage side and upper wing, transforming the handy but slow biplane into a sleek monoplane at the flick of a switch. You wonder if the idea is insane, but after due consideration you decide it may well be the next big thing in aerospace technology.

Somehow, the approval of the chief directorate of the aviation industry was obtained for this idea, and a folding fighter – the IS-1 – was built. Amazingly, for such a seemingly radical machine, it performed excellently. A productionised version dubbed the IS-2 was quickly developed but its monoplane abilities were insufficiently competitive and Nikitin devised the considerably more formidable IS-4. The two wing designs remained basically unchanged, but this is where the similarity ended, as the IS-4 was to be fitted with a bubble canopy, tricycle undercarriage and the M-120 – a 16-cylinder X-configuration engine delivering 1,650 hp. With the M-120 engine, a top speed of 447 mph (719 km/h) was forecast in monoplane configuration, heady stuff indeed for 1941, but a landing speed of merely 66 mph (106 km/h) was projected when transformed into a biplane.

An aircraft offering this astonishing breadth of performance would have been invaluable for the Soviet Air Force, especially early in the war, when their fighters were required to operate from rough fields; the docility and inherent short take-off and landing (STOL) capability of a biplane would have been greatly appreciated. It is also worth pondering what might have been had the design been known to the outside world at the time; the folding fighter concept has obvious potential for carrier-based aircraft, for example. Likewise the inherent liabilities of the type were never to be operationally evaluated – what would happen if the lower wing deployed asymmetrically, for example? Nikitin had designed a lock to prevent this from occurring, but who knows what would happen in combat. Similarly, the undercarriage could not be lowered in monoplane configuration. Were the wing and wheels to stick 'up' for any reason, the resulting forced landing would be highly dangerous and almost certainly result in the loss of the aircraft.

But this was all to remain academic for fate intervened (as it did

'The Soviet Union had been working on rocket-powered research aircraft since the early 1930s, and work on a rocket interceptor, the B1, began in earnest in 1940'

for so many other hopeful Soviet armament projects) in the form of a massive German invasion, curtailing work on promising new aircraft to concentrate on existing types. To be fair, things had already begun to unravel somewhat for the IS-4 when the M-120 engine was cancelled and the lower-powered Mikulin AM-37 (as fitted to the less-than-spectacular MiG-3) had to be substituted as the only alternative inline power unit available. Nonetheless, the IS-4 was apparently flown in the summer of 1941, but records of what flight-testing was done were lost when the design bureau and workshop were evacuated ahead of the advancing German forces.

Despite the recorded completion and flight of the IS-4, I have searched online for nearly 5 whole minutes and not been able to find a single photograph of the complete aircraft. There are three views and an oft-reproduced drawing of the aircraft in its M-120-engined form hurtling skyward in dramatic fashion, but that's about it. Given that every other obscure fighter I can think of has turned up in at least one photograph (even the long-lost PZL.50 Jastrząb), it casts some doubt on the flight claims of this amazing aircraft. Or maybe I just didn't look hard enough. Regardless, whether or not the IS-4 actually flew, its cancellation brought to an end the development of the world's first serious attempt at a variable-geometry-winged fighter, closing the door on a conceptually unique aircraft that appeared to have a great deal of potential. However, the concept of wings that could radically alter their geometry would return with the later 'swing wing' aircraft.

ABOVE: Conceived in the early days of the Cold War, the Article 468 interceptor would require a fast rate of climb to counter contemporary bombers like the US's Boeing B-29. Soviet designer Aleksandr Sergeyevich Moskalev had experimented with delta-wing aircraft since the early 1930s, though the 468 may have been more influenced by the work of the German aerodynamicist Alexander Lippisch. It is not known if stolen Soviet plans aided the design of Roger Ramjet's aircraft.

8. Article 468 (never flown)
'Rocket from the tombs'

No one but the Soviet Union could name things as well without actually naming them. Take the satellite planned to be the first manmade device in space, which was given the mundane and yet somehow awesome moniker 'Object D'. Another example of this minimalist naming policy was the rocket-powered interceptor aircraft, developed by the research institution OKB-2 in the late 1940s, called 'Izdeliya ['Article'] 468'. The 468 was somewhat ambitious for the late 1940s, an era when the major military nations expected the immediate main threat in the near future to be fleets of supersonic bombers penetrating their airspace at high altitude. The Soviet Union had been working on rocket-powered research aircraft since the early 1930s, and work on a rocket interceptor, the B1, began in earnest in 1940. In many ways, the 468 was the culmination of this effort – a slender dart with surprisingly small delta wings and a surprisingly huge tail fin, aided by large fins under the wings that also housed the landing skids.

The Soviet space programme proved there was nothing wrong with its rocket technology. In true Dan Dare fashion, the 468 would take off with the help of a rocket-powered dolly, before using its multi-chamber, four-nozzle liquid rocket motor to climb 72,000 feet (21,946 m) in 2 minutes, guided to its target at up to Mach 2 by radar in the nose. The design was expected to be impressively stable in flight but would have been interesting to land, given that its wing-loading was more than double that of standard contemporary fighters. It's a shame that none of the many pure-rocket interceptors of the late forties and early fifties made it into the air, especially the 468, which made aircraft appearing twenty years later look a bit staid. All that remains of the 468, following its cancellation in 1951, is a wind-tunnel model at the museum of technology at Dubna.

7. Polikarpov I-185 (1941)
'King Rat'

Nikolai Polikarpov's I-185 was an excellent aircraft stymied by engine trouble, politics, timing and outright bad luck, but it should have been the finest fighter that the USSR fielded during the Great Patriotic War. Slightly smaller than the American Grumman Bearcat but weighing 1,900 lb (860 kg) less in normal loaded condition, it had 2,000 hp on tap, was faster than the contemporary Bf 109F at all altitudes up to 20,000 feet (6,000 m) and had immeasurably better handling. It was recommended for immediate production in the autumn of 1942, yet it ended up as merely an also-ran.

The problems began way back in 1937, when Polikarpov's incredibly successful I-16 was fighting in the Spanish Civil War. Republican forces captured a Messerschmitt Bf 109B, which was evaluated thoroughly by a team of Soviet experts. The consensus was that the 109 was inferior in virtually every regard to the latest I-16, the Type 10. While this was true, it was unfortunate that the Soviets failed to envisage the incredible rate of development of the 109; had they captured one of the considerably better 109Es

that were fielded in Spain in the latter stages of the Civil War, it might have encouraged greater urgency in developing a successor to the I-16. As it was, work on an I-16 replacement proceeded in a somewhat leisurely fashion and aimed for rather conservative performance improvement.

The fighter that emerged was named the I-180 and looked very much like a stretched I-16. Development seemed to be going well until December 1938, when the test pilot, Valeri Chkalov, was killed in the prototype. Unfortunately for Polikarpov, Chkalov was a bona fide national hero of immense popularity. While his body lay in state and was visited by all the principal military and civil dignitaries, the NKVD (the infamous People's Commissariat for Internal Affairs) started arresting members of the design team on suspicion of sabotage. It is said that only the personal intervention of Stalin prevented Polikarpov himself being packed off to the Gulag. Work continued on the new fighter, though the programme was somewhat under a cloud. Meanwhile, Chkalov's home town was renamed in his honour and in 1941 a biopic of his life was made, entitled *Red Flyer*.

After Chkalov's death, a major redesign was implemented and the resulting I-180S looked a lot less like the I-16 that had spawned it. Unfortunately for the new fighter, two prototypes were lost in spins in quick succession, one of which resulted in the death of another test pilot, Tomass Susy. Although ten pre-series examples were built during 1940, the performance of the aircraft was tacitly admitted to be less than world class and a further redesign was undertaken. The resulting aircraft was the I-185 and it was

ABOVE: An I-185 of the VVS (Soviet Air Force) in the Second World War.

RIGHT: The early Messerschmitt Bf 109s were considered inferior to the Soviet I-16 Type 10s in almost all regards.

intended to take either the M-90 engine or M-71, both of which offered nearly double the power of the M-88 fitted to the I-180S. Both engines were troubled, and the particularly issue-prone M-90 was abandoned. The M-71 eventually achieved sufficient reliability to power the first I-185 to fly, in February 1942. The aircraft flew beautifully and the M-71 was obviously getting over its teething troubles. When it functioned properly, its performance was spectacular (a speed of 426 mph, or 646 km/h, would ultimately be recorded) and the future finally should have looked rosy for Polikarpov's purposeful fighter.

However, by this time everything had been thrown into chaos by the Germans having invaded and begun their headlong rush towards Moscow. The Soviets needed lots of fighters immediately and didn't have the luxury of waiting for promising prototypes to mature. Unpopular but available fighters were produced in their thousands, and gradual evolution rather than completely new types ultimately yielded the two major Soviet fighter series from Lavochkin and Yakovlev. The I-185 was so good that it refused to die, however. In November 1942, the three existing prototypes were sent to the front to be evaluated under operational conditions. A report by Captain Vasilyaka, commander of the regiment who trialled the aircraft, was

'When it functioned properly, its performance was spectacular'

unambiguously favourable: 'The I-185 outclasses both Soviet and foreign aircraft in level speed. It performs aerobatic manoeuvres easily, rapidly and vigorously. The I-185 is the best current fighter from the point of control simplicity, speed, manoeuvrability (especially in climb), armament and survivability.' A production-standard aircraft was soon completed, but unfortunately the engine failed and the plane crashed. Development continued with the original three prototypes, one of which crashed and killed its pilot after another engine failure, in January 1943. The M-71 engine was rapidly being considered a dead end. Plans to produce the I-185 with the reliable but lower-powered M-82 were eventually abandoned as the M-82 was required for the La-5, an inferior, but good enough, aircraft that, crucially, was already in production. The I-185 programme was formally cancelled in April 1943, finally depriving the Soviet Union of its finest piston-engined fighter. A little over a year later, Nikolai Polikarpov was dead and his design bureau was eventually absorbed into Sukhoi.

TOP LEFT: Valeri Chkalov meets one of the Mario Brothers.

TOP RIGHT: In 1939, Polikarpov was ordered to take a work trip to Germany. While he was away, all his mates fucked him over. His plant director, chief engineer and the design engineer Mikhail Gurevich pitched a new fighter (the I-200), and got the go-ahead to create a new design bureau headed by Artem Mikoyan (whose brother was a senior politician – just saying). On his return, poor Polikarpov found that his design bureau no longer existed and his engineers had been installed at the Mikoyan office. Just goes to show, never go on holiday if you work with knobs.

6. Sukhoi Su-47 'Berkut' (1997)
'Back-to-front to the Future'

In some parallel universe where Salamander Books' *Illustrated Guide to Future Fighters and Combat Aircraft* is an aviation history book, crowds at air shows are wowed by weird-looking fighters performing impossible manoeuvres, with their wings seemingly stuck on back-to-front. In this parallel universe, production versions of the Grumman X-29 and British Aerospace P.1214 rub shoulder pads with Russia's Sukhoi Su-47 'Berkut' – a forward-swept-wing (FSW) experimental heavy fighter from the 1980s. Like shoulder pads, FSWs were briefly fashionable in the 1980s, as they promised enhanced agility, lower take-off and landing distances, and better control at high angles of attack.

While Russia had toyed with a captured Ju 287 bomber after the war and tested their own Tsybin LL-3 in 1948, extreme instability and the strong wings needed meant forward sweep had to wait for fly-by-wire technology (where pilot control inputs are mediated by a computer before reaching the controlled surfaces) and composite materials for designers to be able to create a practical aircraft. Enter Sukhoi, who, in 1983, were given the go-ahead to develop the Su-47 (originally Su- or S-37) demonstrator, which was based on the Flanker family but with fly-by-wire, forward-swept wings and canards.

The Su-47's development was disrupted with the end of the Cold War, and the aircraft didn't get into the air until 1997, a dark time for Russian aviation (though Sukhoi was in a better position than most thanks to Flanker export sales). By this point, technology had moved on, and the Su-47 had been superseded.

While the Berkut's fly-by-wire controls and composite structure undoubtedly fed into Sukhoi's Su-35 and PAK-FA programmes, its radical forward-swept-wings did not. Fly-by-wire and thrust-vectoring enables the Su-35 today to perform jaw-dropping aerobatics without needing canards or FSWs. Stealth, too – where the alignment of edges is the first step in lowering radar cross section (the type's ability to reflect radar signals in the direction of hostile radar receivers) – would also present a unique problem for anyone now designing a FSW fighter. Though only one was made, the Su-47 still looks unbelievable cool. More moderately forward swept wing would later be seen on the Russian KB SAT SR-10 jet-trainer prototype).

'The Su-47's development was disrupted with the end of the Cold War, and the aircraft didn't get into the air until 1997'

5. Sukhoi Su-37 (never flown)
'Крокодил Гена'

As the Cold War was coming to its close, the craze across the fighter houses of Europe was for canard deltas (meaning a triangular wing with the tail at the front as a 'foreplane'). Soviet designers had been studying canard foreplanes on jet fighters since the 1950s, and were reawakened to the idea by both advances in flight-control software and the trend for them in the West (the Soviets' Mikoyan-Gurevich Ye-8 of 1962 was probably the first delta aircraft to use moveable canards as a control surface, predating the IAI Lavi, the British Aerospace EAP and the Dassault Rafale by two decades).

It was at this time, in the late 1980s, that the Sukhoi bureau considered a new ground-attack aircraft that would combine the canard delta configuration with several unusual features. Dubbed the 'Su-37' (this designation was later recycled for a Flanker variant unrelated to this project), this latest Sukhoi development was a single-engined, single-seat fighter.

Having learned from their experience in Afghanistan, the bureau designed the 37 to replace Soviet Aviation's 'Fitters', 'Floggers' and 'Frogfoots' (or is it 'Frogfeet'?). Again echoing trends in Western defence-planning, the Su-37 was intended to combine ground attack and air-to-air capabilities, with an emphasis on the first attribute. Consequently, it had eighteen external hard points able to carry 8.2 tons (8,300 kg) of stores together with an internal 30-mm gun. Of contemporary Western aircraft, only the Tornado was able to lug around more, and it wasn't as good to look at.

To assist the pilot in carrying out the 37's disparate roles, an ambitious avionics package was planned, with multi-mode radar capable of terrain-following and simultaneous tracking

ABOVE: Following the disintegration of the Soviet Union, many military aircraft passed to successor states, as is the case with this notional Su-37 of the Ukrainian Air Force (Повітряні Сили України). This example is armed with the Kh-58 (AS-11 'Kilter') anti-radiation missile.

LEFT: The underside of this manufacturer's model shows ample hardpoints for guided and unguided weapons. The aircraft would have offered a massive increase in destructive effect over the Floggers and Fitters it would have replaced, as well as far greater survivability.

of up to ten targets against background clutter. In addition, an integrated electro-optical system and a defensive aids suite (DAS) were developed, technologies found today on the Lockheed Martin F-35, and unlike the F-35, the 37 was fitted with 1,764 lb (800 kg) of armour plate to protect the pilot and other sensitive points. Alas, with the ending of the Cold War, funding for this supersonic Shturmovik never materialised, and instead we enthusiasts of Russian metal must be content with endless, tedious Flanker derivatives.

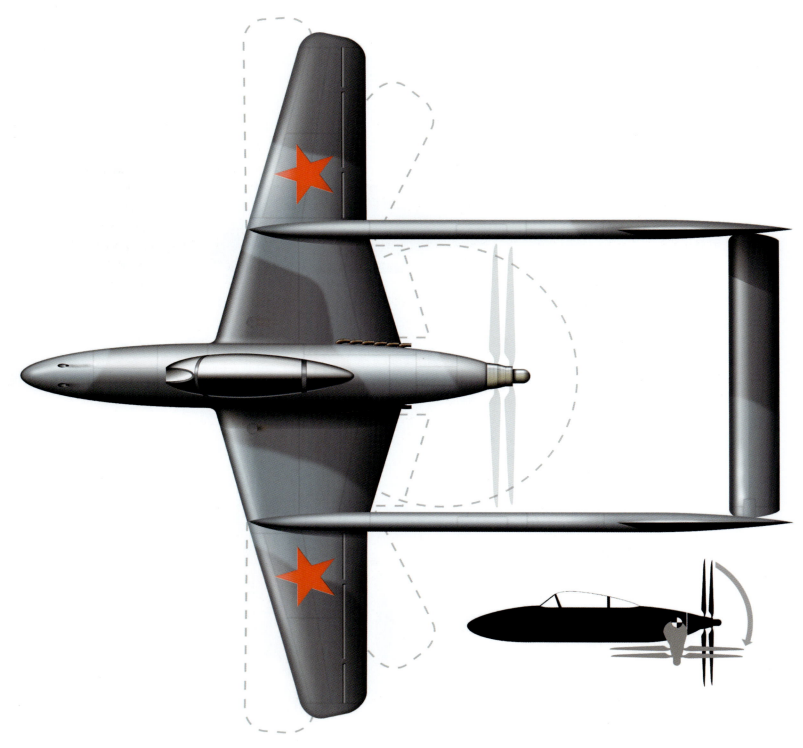

4. Shulikov 'Sh-1' Fighter (1943)
'Машенька'

ABOVE: In every sense an utterly radical design, this plan view of the Sh-1 shows the range of wing sweep. The side view demonstrates the tilting prop required for vertical (or more likely, short) take-offs.

The concept of vectored thrust, used by the Harrier, is based on an idea from the French designer Michel Wibault (echoing some similar but not identical concepts that Britain's Frank Whittle considered). However, the idea of a 'four-poster', vectored-thrust fighter with vertical and take-off landing (VTOL) was put forward even earlier, by the Soviet designer Konstantin Vladimirovich Shulikov. Prior to this, Shulikov came up with a very clever twin-boom pusher tilt-rotor VTOL fighter design. The resulting aircraft would have been fast, and armed with at least four cannon or machine guns. With its clean, streamlined form, tricycle landing gear and unobstructed pilot's view, it had much going for it, but as Shulikov himself realised, it was unlikely to have achieved purely vertical take-off with the engines of the time. Did I mention it also had a variable-geometry outer wing section?

3. Grokhovsky G-38 (never flown)
'Heavy Metal Monster'

In the mid-1930s, the concept of the 'cruiser fighter', or 'Zerstörer', was very popular in design and planning circles. The Grokhovsky G-38 was one of many examples of this class of fighter that never left the drawing board. It was a twin-boom, multi-seat heavy fighter comparable in concept to the Dutch Fokker G.1 or American Lockheed P-58 'Chain Lightning'. The G-38, however, was remarkable in a number of respects, most significant of which was the execution of the twin-boom concept. The Fokker and the Lockheed were large, bulky, even clumsy aircraft, as was the original take on the G-38. When Grokhovsky hired the young engineer Pavel Ivensen to work on the project, however, the aircraft was transformed into something rather exciting. Ivensen started with a clean sheet. The new G-38 was tiny for a three-seat aircraft, with a wingspan of 44 feet (13.4 m) – compared with 53 feet (16 m) for the P-38 and 56 feet (17 m) for the Fokker G.1 –

and boasted ultra-neat packaging. The crew were contained in a torpedo-shaped pod faired into the broad wing centre-section, and the two Gnome et Rhône radial engines tapered to super-slender booms. It had an incredibly low frontal area for an aircraft of its class and a high wing-loading for the time, and it's safe to say that it would have been fast. Most remarkable of all was the fact that the preliminary designs were approved in 1934, making the highly modern-looking G-38 contemporary with the Hawker Hurricane and Curtiss P-36. Had it not been cancelled (for 'unknown reasons', around the time of the major Stalinist purges), it is intriguing to consider what this aircraft might have done for the otherwise lacklustre heavy-fighter class. In 1935, Ivensen himself fell victim to one of Stalin's purges, and he was later sent to a labour camp dedicated to aviation engineering, where he had a hand in designing the Tupolev Tu-2.

BELOW: A G-38 heavy fighter of the Soviet Air Force in 1941. Had this aircraft actually served, it would have been thrown into the desperate defence of the Soviet Union, where it would have been an extremely useful addition to the beleaguered air force.

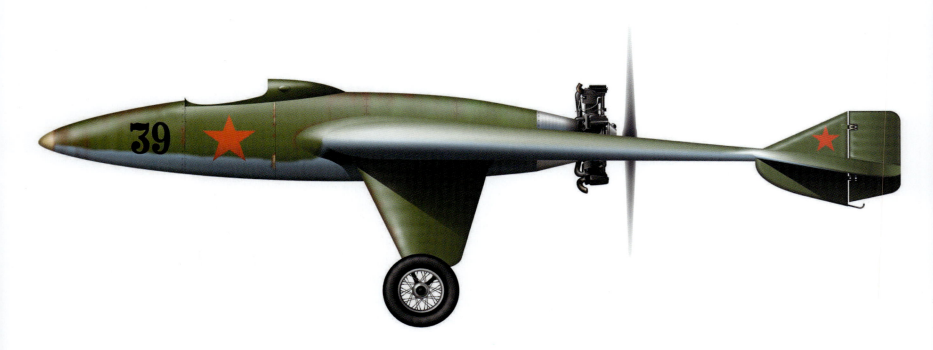

2. Grokhovsky-39 (never flown)
'Ram and probe'

ABOVE: The G-39 was an innovative, and alarming, design. This is how it might have appeared in VVS service.

BELOW: A notional G-39, with its nose rammer and wire weapon fitted. The wire-cutter was intended to slice through the bracing or wings of enemy aircraft. An almost suicidal bravery would have been required from its pilots, a quality not lacking in Soviet pilots of the time.

On 8 September 1914, the Imperial Russian Air Service pilot Pyotr Nesterov performed the first aerial-ramming aircraft attack, using his aircraft itself as an offensive weapon. Though very dangerous, the use of ramming as a last-ditch tactic proved popular with Soviet pilots, and in 1932 the Soviet Air Force began a classified project to produce a purpose-built ramming fighter. This effort, dubbed Object 'Taran' ('battering ram'), considered various manned and unmanned solutions before settling on Grokhovsky's G-39 project. Grokhovsky was a highly skilled pilot,

aircraft designer and inventor; he created the world's first cotton parachutes and designed items as varied as cargo containers for airborne troops, rocket artillery, armoured hovercraft and even a weaponised snowmobile.

The G-39 design detailed a monoplane pusher with rudders on the outer sections of the wing instead of on a conventional tail unit. The aircraft's most unusual feature, though, was its weapon: two steel wires running from a boom on the nose to the each of the wingtips, intended to slice through enemy aircraft.

'The G-39 design detailed a monoplane pusher with rudders on the outer sections of the wing instead of on a conventional tail unit'

In case the wires snapped, the wing's leading edges were made exceptionally strong. The exceptionally brave (or unfortunate) G-39 pilots would have had a degree of protection from a retractable bullet-proof windscreen. This extremely strange machine was readied for flight in 1935 but refused to take off – its 100 hp engine made it woefully underpowered. Work on the G-39 was discontinued, and like many others, Grokhovsky was crushed by Stalin's brutal state – he was arrested in 1942 and died in prison four years later.

TOP LEFT: Katya Grokhovsky, *Untitled Heroic*, 2011. It is not known whether the artist Katya Grokhovsky is a descendant of the aircraft designer. (Well, to be honest, we do know the answer as we emailed her to ask. She is not related.)

1. Mikoyan-Gurevich Ye-150 series (1959)
'Soviet hotrods'

The Ye-150 series were wildly-high-performance heavy interceptors. They could out-drag and out-climb any fighter in the world, and they also looked exceptionally mean. Despite taking its first flight as long ago as 1959, the Ye-150 could reach an astonishing Mach 2.65 (some sources claim even higher speeds) and could ascend to altitudes above 69,000 feet (21,031 m). (Remarkably, all of this was achieved with the same installed thrust as today's rather more pedestrian Saab JAS 39 Gripen.) The series of four experimental fighter-prototypes was built in an effort to create a new, highly automated fighter to defend the Soviet Union against a proliferating Western threat (including supersonic bombers like the B-58, then in development). To catch and destroy these fast, high-flying intruders, the interceptor was to be automatically steered under the guidance of ground radars before engaging its own cutting-edge detection and weapons systems.

But it was a case of too much too soon; the ferociously exacting requirements for the electronics, missiles and powerplant were too demanding, and each suffered severe delays and development problems. What could have been the best interceptor in the world was ultimately cancelled in 1962.

'Despite taking its first flight as long ago as 1959, the Ye-150 could reach an astonishing Mach 2.65'

OVERRATED
AIRCRAFT

Oasis, *The Godfather*, champagne – many things in life are overrated. Popular opinion will hold that they're the outstanding examples of their kind, but popular opinion is merely the collective braying of the uneducated hordes. To elevate you above the uneducated hordes, the following list contains ten of the most overrated military aircraft, allowing you to display a veneer of sophistication when they crop up in conversation. To be clear, most of these aircraft aren't bad; generally they've at least displayed some level of basic competence, but this has been over-inflated in the popular imagination to an unwarranted degree. The Bruce Springsteens of aviation, if you will.

To avoid filling this entire list with flights of fancy, cancelled projects don't count. This saves you, the reader, from my multi-volume rant about the TSR.2 having fewer flying hours than the X-35 did when it was selected to be a Joint Strike Fighter. The list also pretty much comprises only combat aircraft, as it's hard to think of any overrated cargo haulers. Or helicopters. Or Blackburn products.

National chauvinism frequently plays a part in the overrating of aircraft, or of anything else, for that matter. For this reason, the list would appear to be biased against British aircraft, because those are the ones the author has most often heard being praised while thinking, 'Steady on, they aren't that good.' Plus, let's face it, most other countries' aircraft are average at best.

McDonnell Douglas F-4 Phantom II (1958)
'Smoking Satan'

Hush-Kit's top Cold War carrier-borne combat aircraft is the F-4 Phantom, probably the first jet aircraft to succeed at being multirole and so good the United States Air Force swallowed its pride and bought a few. Thousand. Contrary to popular opinion, and some dialogue in *Top Gun*, it also achieved a respectable kill-ratio in the Vietnam War compared with smaller, more agile opponents. However, it wasn't without its problems.

Its design as a solely missile-armed, all-weather interceptor proved to be overly ambitious when put to the test in Vietnam, where the performance of the early Sidewinder and Sparrow missiles were sub-optimal. Initially this led to crews' ripple-firing missiles in an attempt to gain a hit. But with the close infighting that developed due to the rules of engagement that were in force, even this wasn't guaranteed to work. Indeed, it was not uncommon for a Phantom to find itself too close to the enemy to fire even the short-ranged infrared-seeking missiles. Initially this led to the fitting of gun pods, the accuracy of which was variable, especially when attached to aircraft subject to catapult launches. Ultimately an internal gun was fitted on the aircraft from the E-model onwards, a solution that required the shuffling around of internal components to maintain the centre of gravity and the removal of the ram air turbine that supplied hydraulic pressure in the event of engine failures.

To add to the woes in Vietnam, the J79 engines used by the Phantom produced copious amounts of black smoke, providing a convenient pointer towards the aircraft for any enemy fighters or anti-aircraft batteries. Still, at least it only took two decades to fix that issue, with the J79-10A fitted to the F-4S in 1977 – two years after the end of the war in Vietnam. Having two of the J-79s also meant the 'Toom drank like a furloughed bachelor during lockdown, getting through about five times as much fuel as a Harrier or Corsair just to get airborne.

Having just about got the F-4 sorted, the world's air forces started replacing it as newer designs were introduced that fixed its various shortcomings. (The Royal Navy was no different, though they chose the Sea Harrier over the F-4 because it could operate from their new compact carrier, not because it was a better aircraft). Despite this, the Phantom remains in service today, with a variety of air arms in place in 2021, sixty years after the formation of the first US Navy squadron to operate the aircraft. Iran, South Korea, Greece and Turkey are still nominally operating McDonnell Douglas' finest, the last two countries presumably refusing to retire their aircraft until the other does.

The Phantom was a good aircraft at its peak, although it was inevitably compromised when compared to more specialist airframes such as the A-6 or F-8. But this peak took place broadly during the late sixties through to the early eighties; for the other two thirds of its life it's either been struggling through development woes or stumbling around in the early hours trying to find the toilet as bladder control starts to become an issue.

Don't say: Confusingly mixed messaging from the clowns at Hush-Kit.

Do say: Good for its time, but increasingly outclassed during the latter four decades of its service.

LEFT: The French term *jolie laide* refers to someone who is both ugly and attractive, and this is certainly applicable to the ugly-beautiful F-4 Phantom II, a mess of ungainly lines that somehow add up to an utterly impressive overall appearance.

Heinkel He 113
'The Super Fighter'

It's fair to say the first half of 1940 hadn't gone well for Britain and her allies; having lost the battles of Norway and France, everything relied on winning the upcoming home match to stay in the championship. Worryingly for Fighter Command, who would bear the brunt of the impending battle and had only just held their own against the Luftwaffe's current fighters, the Bf 109 and Bf 110. It was obvious from intelligence reports that the far-superior He 113 would be a completely different proposition. With a top speed of 390 mph (628 km/h) – some 35 mph (56 km/h) faster than the Spitfire – and a fuel-injected engine allowing negative-G flight, the German plane would be able to run rings around the RAF's best. Perhaps more worryingly, where the British aircraft were armed with eight or sometimes twelve .303-calibre machine guns, photos indicated the Germans had managed to incorporate three 20-mm cannon into their fighter. For those not up on the technicalities of fighter weaponry, machine guns fire inert lumps of lead, 0.303 inches (0.76 cm) in diameter in this case. Cannon, on the other hand, fire mini projectiles that incorporate explosives that detonate on contact, which makes much more of a mess. The Allies were working on arming their aircraft with cannon but these wouldn't be widely adopted until the following year due to difficulties developing them and achieving performance hit, resulting from their greater mass.

Everything indicated that the He 113 was going to be a tough customer, and so it would prove. The first encounter seems to have occurred while Hurricanes were covering the Dunkirk evacuation, when they were bounced by the 'Super Fighters' while themselves preparing to attack a group of He 111 bombers. This pattern would repeat itself throughout the subsequent, rather predictably named Battle of Britain, where flights of He 113 would strike from high level just as the attacking fighters were about to engage bombers. Later, they would use their great speed to carry out lightning raids on ground targets, the first on 18 August 1940, when they destroyed a Hurricane and seven Spitfires at RAF Manston for no losses. Indeed, throughout the battle there were no confirmed kills of the He 113, only a handful of probables.

The only relief for the Allies was that the Germans seemed to have limited numbers of the aircraft available. It was speculated that this was due to difficulties operating what was obviously a highly advanced aircraft from muddy fields in France. In reality it was because they hadn't actually built any He 113s. The whole thing was a ruse by the Nazis' head of propaganda Joseph Goebbels and the Luftwaffe, using repainted He 100 prototypes to convince their enemies they were far ahead of them in aircraft development. The He 100 had set a world air-speed record shortly before the war so this was a plausible basis for a new fighter, but for a variety of reasons it hadn't been selected as such by the Luftwaffe. The 113 designation was chosen in an attempt to play up to the stereotypical image of German methodical predictability, being a logical follow on to the He 112 that had seen limited service. That the ruse worked can be demonstrated by the willingness of Allied pilots to report higher-flying, faster aircraft as He 113s, when in reality they'd just been Bf 109s using a height advantage to gain speed. More crucially, as the fighter's reputation grew, the Allies would become increasingly wary of engaging any formation they believed to contain He 113s.

To say the He 113 was overrated is probably understating things. Fear of its ability allegedly factored in Dowding's decision not to deploy Spitfires to France, when in reality the He 113 was a prototype cosplaying as an end-of-level baddy.

Don't say: What?

Do say: The kind of information warfare that Alastair Campbell would be proud of.

BELOW: As a deterrent, the He 113 was undoubtedly effective. It was a hugely cost-effective deterrent as no money was spent on it (other than for some paint) and none were built.

Avro Vulcan (1952)
'Hero of Operation Slack Fuck'

Say V-Bombers and, if you're basic, your first thought is the Avro Vulcan. And why not? It's a moderately attractive cranked delta with four Olympus engines. Its greatest claim to fame is, of course, the Black Buck raids during the Falklands conflict of 1982, which saw pretty much the entire surviving V-force execute possibly the most complicated refuelling plan in history to hit a runway. With a bomb. Followed shortly after by the retirement of most of the Vulcan fleet, at the end of the year. The kind of thing someone should write a book about.

So, job done, Vulcan: greatest V-bomber, if not strategic bomber, of all time, right? Well, frankly, no. It's certainly in the top three V-bombers, beating the Short Sperrin by actually entering service, and the Vickers Valiant by not falling apart after ten years of pootling around in the sky. Although, to be fair, the Valiant did have the distinction of actually dropping a nuclear bomb and it beat the Vulcan to conventional bombing missions by twenty-six years during the Suez Crisis. (It managed more than one hit on the runway as well.)

Still, if you're a Top Trumps kind of aircraft fan, you'll be interested to know it could at least carry the same bomb load as the Valiant but slightly faster. It fails utterly in comparison to the other V-bomber, though. The Handley Page Victor could carry fourteen more 1,000 lb bombs than its sisters in the V-force, making a total of 35,000 lb, or half the payload of a B-52. (The Vulcan was tested with a load-out of thirty 1,000-lb bombs, but this was never put into practice in service and probably wasn't cleared for operational use in any case.) It could also go further and, when in a shallow dive, break the sound barrier.

The only real problem with the Victor was the manufacturer, Handley Page's chairman being less keen on the government's ideas for manufacturer consolidation than the company were. That,

along with potential issues fitting the Skybolt missile, meant only a handful of Victor B2s were ordered compared to an extravagance of Vulcans. Consequently, when Skybolt was cancelled, the greater number of Vulcans meant they were a shoo-in for the bomber role, while the Victors became tankers.

Not a bad aircraft overall, the Vulcan's reputation today seems to rest on its role in Black Buck rather than its overall capability. Plus, it appears in the James Bond flick *Thunderball*, which doesn't hurt.

Don't say – A triumph of British design and engineering.

Do say – The Victor could carry more, further. What else do you want in a bomber?

ABOVE: The disproportionate popularity of the Vulcan is largely the result of its good looks and noise. The Vulcan was slightly faster and performed better at high altitude than rival bombers, but the small speed advantage came at a great cost.

Mitsubishi A6M Zero (1939)
'Zero some'

To say that the Mitsubishi Zero came as an unpleasant surprise to the Allied forces during December 1941 would be an understatement. Nimble, fast and long-ranged, it was everything you'd want in a carrier fighter. Okay, slightly more armour and radio aids would be nice, but their absence didn't stop it racking up an impressive kill-rate during the sudden expansion of the Greater East Asia Co-Prosperity Sphere into the Indian and Pacific Oceans. During raids on the only 'civilised' bit of northern Australia, Darwin, they easily dealt with the Spitfire Vs that had been rushed there for its defence. This, despite the defenders being flown by experienced Battle of Britain veterans.

The evidence for the Zero's reputation seems pretty clear-cut, then, cutting a swathe before it even when opposed by the 'greatest fighter of all time' flown by the best of the best – or, at least, the best of the RAF. Mitsubishi had obviously created some sort of uber-fighter, probably either with the direct help of the Germans or inspired by an obscure Gloster design that never got ordered. (The Gloster F.5/34. Google it – it's pretty similar.) Because the alternative would be that the Zero was an average fighter with strengths and weaknesses and that the Japanese were just producing better pilots than the Allies. Which would be incredibly inconvenient given all the Western intelligence assessments and propaganda that had asserted the Japanese were producing inferior knock-offs of Western aircraft and were themselves physically inferior, and particularly inconvenient if you'd then based your defence policy on those assumptions and left a variety of obsolete aircraft to defend your key outpost in the region. Still, at least the loss of

Singapore put an end to any conceit of racial superiority in the British populace.

The Zero wasn't a bad aircraft; it was very manoeuvrable, and had decent armament and excellent low-speed handling. However, it wasn't particularly fast, and its critical altitude, where the supercharger can no longer compensate for the depredation of altitude on engine performance, was a relatively low 16,000 feet (5,000 m) – useful for naval air combat but much lower than that of the Spitfire and other fighters optimised for the European theatre. Its main advantage was in the cockpit, where the pilots sat after undergoing what at the time was the longest training course in the world – this allowed them to make the most of their aircraft and to drag Allied pilots into combat on their terms. Meanwhile, the Australian Spitfire pilots insisted on using tactics that wouldn't have cut the mustard in Europe, by getting into turning fights with one of the few aircraft that could out-turn them. Indeed, the score could probably have been reversed if they'd used high-speed dives to make slashing attacks on the Zeros before climbing away to position for a follow-up.

As a fighter the Zero was a good aircraft. However, to compensate for the shocking performance of the Allies against what they'd been told was an inferior enemy, it had to become a great one.

Don't say: It swept a wave of terror across the Pacific.

Do say: Made into a bogey man by the Allies to cover up their pre-war intelligence failings.

ABOVE: The A6M had an unbeatable rate of turn and excellent range, but could be effectively countered with appropriate tactics. Though brilliantly designed and constructed, it was a rather mediocre performer and suffered when facing the world's best fighters.

Boeing B-17 Flying Fortress (1935)
'Captain Mediocre'

Ask the average man in the street to name a Second World War US Army Air Force (USAAF) bomber and he'll probably ask you to stand two metres away and wear a face mask. Repeat the experiment enough times, however, and you'll soon realise that the Flying Fortress is probably the most famous American bomber of the war. Capable of delivering 4,500lbs of high explosives to Berlin while fighting off hordes of Fw 190s and Bf 109s with an ever-increasing battery of .50-calibre machine guns, the B-17 made a major contribution to attempts to pummel Germany into submission.

Impressive as that sounds, it was also broadly what the Mosquito could do on half the engines, only two crew yet going about 130mph faster. Meanwhile, the USA's other heavy, Consolidated Aircraft's B-24 Liberator, could fly further, faster and with a greater bomb-load. It was also the most-produced bomber in history, with 18,482 to the B-17's mere 12,731. (The Ju 88 making second place with 15,183 built, fact-perverts.)

So, why the love for what was at best the equal of its peers? Mostly good PR. Even before the USA was bombed into entering the war against fascism, the B-17 was getting publicity for its work with the RAF. Ironically, this was mostly with Coastal Command, the early models being considered unsuitable for the aircraft's intended primary role after initial trials over occupied Germany had achieved at best mixed results.

With the late entrants to the war doing their best to play catch-up, the US Army Air Force were keen to push the message that they were taking the fight deep into Nazi Germany. Until mid-1943, this meant lots of reports of Flying Fortresses, cementing its place in the public's heart before the B-24 had got down to business. If this wasn't enough, just as the Liberator was deploying in May '43, the first B-17 completed a tour of twenty-five missions. The following week the *Memphis Belle* completed its twenty-fifth mission to a blitz of publicity, including the 1944 release of a documentary broadly documenting that final mission. In a further insult to, well, everyone, the 1990 film *Memphis Belle* fictionalised the making of the documentary, added to the profile of the B-17 and gave Harry Connick Jr an acting career.

Essentially, then, the Flying Fortress was an all-right bomber with a great PR department – the Adam Sandler of strategic bombing. Seriously, how is he still getting roles?

Don't say: The bomber that won the war.

Do say: The Liberator did the work, the B-17 got the glory.

BELOW: B-17 42-97272, Duchess' Daughter, crash-landed in 1944. The B-17 was slower than the B-26 and carried less ammunition for its defensive guns.

'With the late entrants to the war doing their best to play catch-up, the US Army Air Force were keen to push the message that they were taking the fight deep into Nazi Germany'

Harrier (1967)
'Vectored trust'

The 1960s were a time of great experimentation as the world's aircraft manufacturers attempted to produce viable aircraft with vertical and short take-off and landing (VSTOL). British aviation enthusiasts take a considerable degree of pride in the emergence of the Harrier as the only successful aircraft from this era of sometimes crazed tinkering. (No, the Yak-38 Forger *doesn't* count as successful.) This pride is enhanced by the Harrier's ability to inflict wide-area tinnitus via air-show performances.

But you do have to question whether it was all worth it. The genius of the Harrier was in doing away with a separate lift engine by using the front stage of a turbofan to provide thrust forward of the centre of gravity. The hot stage does the same aft of it; a bit of shuffling around with the aircraft's mass and, *voilà*, you can hover. However, it does mean you need a turbofan with a reasonable bypass ratio, which is the sort of thing you normally see on a business jet. It also makes engine maintenance something of a challenge, the wing having to come off if you want to change it. Plus, there's the usual problem with any powered-lift aircraft using full thrust for a vertical landing, which is the increased fatigue and vibration load, decreasing airframe life.

The proof of the pudding is in the eating, and it's notable that the Royal Air Force was the only land-based operator of the Harrier. All the other Harrier variants were sold to people who wanted to fly from ships, the US Marine Corps, the Spanish Navy, the Royal Navy,

the Italian Navy, etc. Indeed, the major operation that the RAF's second-generation Harriers were involved with was Afghanistan, where their VSTOL capability was well used, operating from a runway otherwise suitable only for Tornadoes.

Another area of misplaced pride in the Harrier is the assumption that it's somehow better than the F-35 because it doesn't have a lift fan and the UK 'should have just made a better Harrier rather than wasting money on the F-35'. This overlooks the fact that the Harrier's lift fan is essentially the first stage of the Pegasus, is always engaged and creates an increasing amount of intake drag as you approach Mach 1. Never mind that for a turbofan producing a similar level of thrust to the F-35's engine, you'd be have to be looking at something that's usually hung off an A320. Good luck building a fighter around that. In any case, the Pegasus configuration makes for some awkward packaging decisions when designing your aircraft: look at the P.1214 to see some of the workarounds needed to use it on an advanced short take-off and vertical landing (STOVL) project.

The Harrier, then: a really successful technology demonstrator that probably degraded the UK's overall defence capability. Still, handy if you're going to cheap-out on your aircraft carrier.

Don't say: A triumph of British design and engineering.

Do say: Pardon? I can't hear you over the sound of the Harrier.

ABOVE: It was planned that in times of war the RAF's Harrier force would be based at camouflaged locations in woods, at petrol stations or anywhere that could accommodate them. In reality, off-site basing is notoriously difficult to support, though it would take aircraft away from vulnerable known bases.

Supermarine Spitfire (1936)
'The Shaming of the Shrew'

If you're British and aren't particularly interested in aircraft, the Spitfire is your favourite. It's probably also what you call every camouflaged aeroplane with a propeller that you see. This helps the aviation connoisseur to avoid you.

Not that the Spitfire was a bad aircraft. As a short-range interceptor it was exactly the sort of thing you'd want if, say, you were planning on defending an island against an aerial onslaught. Once that unpleasantness is out of the way, however, you really need something with more range. Despite photo-reconnaissance units and the US Army Air Force proving you could usefully increase the Spitfire's fuel-load to almost equal the range of the Mustang (by putting tanks all over the aircraft rather than just one behind the engine), the RAF proved resistant to the idea. This made it difficult to provide an escort to any missions going further than, say, the beaches of Pas-de-Calais.

Nor did this limited range gain the Spitfire much in the way of performance: thanks to a more aerodynamic form, with broadly similar Merlin installations the P-51C was 7 per cent faster than the Spitfire Mk.IX with 10 per cent better fuel consumption. This, despite the Mustang being 20 per cent heavier. The Spitfire was faster-climbing, though, which did at least allow it to spend its limited time airborne at a decent altitude.

Long-range escort not being an option, attempts were also made at dive-bombing, where the Spitfire would build up speed too quickly. This would have been less of a problem if it hadn't lead to the ailerons detaching if unchecked by the pilot. The light build

'The less said about the attempts to make the Spitfire seaworthy, the better'

of the aircraft also meant it could carry only half the bomb-load of its contemporaries – 1,000 lb (450 kg) to the P-51's and P-47's 2,000 lb (907 kg).

It was, however, unrivalled in air-to-air combat – apart from, say, against the Zero, or when operated by the Soviets, who relegated it to the rear with the PVO's air-defence forces, favouring the Bell P-39 Airacobra for the VVS busy engaging the Luftwaffe over the battlefield. That's right – they preferred the Airacobra, something virtually no one has heard of, to the Spitfire.

The less said about the attempts to make the Spitfire seaworthy, the better; suffice to say the undercarriage was never really up to the job, even with multiple upgrades. By the time it got to the Korean War, the Seafire model was stretching the original design so much that the fuselage would wrinkle from conducting deck landings. A bit of a design flaw in a carrier aircraft. The Battle of Britain gave the Spitfire great PR, but this has overshadowed its later, perfectly average performance.

Don't say: A triumph of British design and engineering.

Do say: The sort of dead-horse flogging that saw the original Mini in production for forty years.

BELOW, FROM TOP: An RAF Spitfire Mk.XVI; an FAA Seafire LF Mk.III (894 Sqn, HMS *Indefatigable*); an Israeli Spitfire Mk.IV (109 Sqn).

A BANGING
HISTORY OF
THE BANGSEAT

A flight can go wrong very quickly. Sometimes you need to leave the cockpit extremely swiftly to avoid becoming a burned human pâté of gristle and teeth. Parachutes had been around for a while when the aeroplane entered the scene in the twentieth century, but their development hadn't been an entirely smooth one.

One of the earliest attempts at a proto-parachute took place in the year 852, in Córdoba, Spain. A Moorish polymath-poet successfully transmuted altitude into acute back pain; he leapt from a tower, hoping the cloak of feathers he was wearing would ease his fall. He was disappointed. Many great (and not so great) minds experimented with the idea of the parachute for the next few centuries, including a certain Italian named after a Teenage Mutant Ninja Turtle. It took a long time – and quite a few dead monks, madmen and polymaths – to perfect it.

The inventor Louis-Sébastien Lenormand made the first successful public jump in 1783 in France, the same year that saw the start of human flight. Jumps from air balloons soon followed, and the parachute was further refined over the next 120 years or so. But the careful unstrapping and bodily manoeuvring required to depart a crashing aeroplane was dangerously slow, and often extremely perilous. A better solution was sought, most obvious of which was to propel the pilot out from the aircraft safely and at great speed. The search began for a way to do this.

In 1910, a British university professor demonstrated a cartridge-fired bungee to eject a parachute from an aircraft (whether it was faster than the badly tethered bungee that inevitably whips off a family-car roof rack is unknown). In 1912, Adolf Odkolek, inventor of the Hotchkiss machine gun and several other weapon technologies, demonstrated a cartridge-fired spreader gun to deploy a parachute. Two years later, George Prensiel, an engineer working at the London Aerodrome, Hendon, put his compressed-air parachute-extraction system to the test, firing it from the rear of a moving car. By August he was carrying out live tests above the aerodrome. In 1916, Everard Calthrop, a railway engineer and famous breeder of Arab horses, invented and marketed a 'British Parachute' and the 'Guardian Angel' parachute. In the same year, he patented the first ejection seat (which used compressed air) but it did not go into service.

The Great War was murderously dangerous for airmen. By the end of it, military pilots had a one-in-four chance of being killed in action (but it should be noted that in 1917 more men were lost in the training schools than on battle fronts). Many British pilots were happy to buy their own parachutes but were denied the right to use them by somewhat psychopathic officials, who thought that this would encourage the pilots to avoid combat.

Work on a fast-escape system persisted. In 1922, Edward Scheemaker patented his spring-loaded Concertina EjectaSeat in the US, making it the first US patent issued for a compressed-gas ejection seat. In 1928, Romanian inventor Anastase Dragomir apparently perfected a 'catapultable cockpit'. He and his colleague, the engineer Tănase Dobrescu, applied for a French patent and in 1929 Dragomir began construction of his device. The invention was tested in a Farman aeroplane on 28 August 1929, reportedly with success. In the UK in 1930, RAF Flying Officer Tony Dudgeon, alarmed by the perils of the high speed of contemporary aircraft, proposed a new design of an 'escape seat', with two telescopic tubes using highly compressed springs. This idea was submitted to the British Air Ministry, who weren't interested. British innovation in aircrew escape would have to wait until the fabulously named Operation Lusty.

The German Research Institute for Aviation (DVL), however, experimented with ejection systems with great scientific rigour in 1939. Physiological testing devices measured the forces of gravity and acceleration on the human body. These tests determined how well a human body could withstand G-force onset of more than 20G for a tenth of a second – the speed and force the DVL believed a future ejection system should achieve. In 1941, the DVL carried out the first in-flight ejection test using a test mannequin fired from the rear gunner's position in the Ju 87. In the same year, Saab of Sweden applied for their first ejection seat patent, and in 1942 their first airborne ejection tests using compressed air took place aboard a Saab B 17 aircraft (not to be confused with the American Flying Fortress) using a test mannequin. The first recorded case of a live emergency ejection was that of Heinkel test pilot Helmut Schenk, from an He 280 on 13 January 1942, using a compressed-air powered Heinkel seat (later, Heinkel seats used an explosive charge developed for use in the He 162). By late 1942, all German experimental aircraft had some form of Heinkel ejection seat. The He 280 and He 219 were the first production aircraft with ejection seats included in the design, and it is estimated that some sixty emergency ejections were made in the war by Luftwaffe airmen, though it is not known how many were successful.

In 1944, the Allies' Operation Lusty was launched to collect all useful German research, along with some rather unsavoury

OPPOSITE: Capt. Christopher Stricklin ejects from a stricken USAF Thunderbirds F-16 seconds before it crashes during a 2003 air display. The ACES II ejection seat undoubtedly saved Stricklin's life.

> '**In 1945, Lieutenant Rudolf Schmitt successfully ejected from a Heinkel He 162. (If you've seen a Heinkel He 162 up close, you wouldn't blame him.)**'

TOP LEFT: The B-58 Hustler's unusual escape pod required extensive testing, some of which involved an unlucky bear named Yogi.

TOP RIGHT: In testing seats, rocket-powered sleds are often used to recreate the conditions of an ejection at low level.

scientists. One of these was Siegfried Ruff, a German physician who served as director of the Aviation Medicine Department at the DVL and had been acquitted of accusations of earlier lethal human experimentation at Dachau concentration camp. After the war, the US Army Air Force hired Ruff to conduct experiments measuring human exposure to high altitude. Meanwhile, James Martin of the Martin-Baker Aircraft Company (which had failed to interest the RAF in its excellent fighter aircraft) was invited by the Ministry of Aircraft Production to investigate the practicability of assisted escape systems for fighter pilots.

In 1945, Lieutenant Rudolf Schmitt successfully ejected from a Heinkel He 162. (If you've seen a Heinkel He 162 up close, you wouldn't blame him.) Then, in 1946, Lieutenant Bengt-Olov Johansson became the first Swedish pilot to eject in a genuine emergency (from a Saab J 21A-1 aircraft) after a mid-air collision. On 24 June the same year, an in-flight ejection of a mannequin

from a Meteor Mk.III was made by Martin-Baker in the UK, who then carried out a successful live human ejection on 24 July: Bernard Lynch (who had led the development of this Mk.1 seat) ejected himself from a Meteor at 320 mph (515 km/h) at an altitude of 8,000 feet (2,438 m). Lynch again ejected himself, on 19 August 1947, this time at 12,000 feet (3,658 m) at 420 mph (676 km/h). This successfully demonstrated the smoothness of Martin-Baker's new ejection-gun approach and the success of their face screen and the stabilising drogue features, plus Lynch's absolute stone-cold chutzpah. On 30 May 1949, while flight-testing the Armstrong Whitworth A.W.52 'flying wing' aircraft, test pilot 'Jo' Lancaster experienced severe pitch oscillations in a dive and feared the aircraft would disintegrate. He ejected, employing the first emergency ejection using a Martin-Baker seat.

In the US, the Weber company became one of the largest producers of ejection seats, with its seats fitted in the B-47, B-52, F-101 Voodoo, F-102 Delta Dagger and F-106 Delta Dart —over 3,000 seats manufactured in all. Another US aerospace company, Stanley Aviation, founded by the first American jet pilot Robert Morris Stanley, was awarded its first ejection seat contract in 1954 and became responsible for the downward firing seats in the B-47, RB-47, XB-52, RB-66 Destroyer and F-104A Starfighter. Later that year, on 13 October, the first successful underwater ejection took place from a Royal Navy Westland Wyvern. In 1961, the first live 'zero-zero' (zero altitude and zero airspeed)

test ejection was made by William 'Doddy' Hay in the UK with Martin-Baker. Zero-zero technology uses small rockets to propel the seat upward to a sufficiently safe altitude, then an explosive charge to open the parachute canopy particularly quickly.

In 1962, a live drugged bear was used to test the capsule escape system of the B-58 Hustler. Ejected at 35,000 feet (10,668 m) at 870 mph (1,400 km/h), the bear (known as Yogi) survived the incident, only to then be euthanised to have his internal organs inspected. From this, bears learned that humans are never to be trusted. Since the start of human-crewed Vostok spaceflights in 1961, cosmonauts had used an ejection capsule for standard landings. (Before this, they were used by dogs.) If we are going to count these Vostok landings as 'ejections', it means the very first female ejection was probably by Valentina Tereshkova at the end of the Vostok 6 mission in 1963. (As an aside, following a cosmonaut tradition, she had pissed on the tyre of the bus that had brought her to the launchpad at the beginning of the mission.) The Soviets were worried about how intentional this rather dramatic landing technique looked to international observers, and it was not publicly revealed until the early 1990s.

In 1966, two Lockheed M-21 crew members ejected at Mach 3.25 at an altitude of 80,000 feet (24,000 m) after an accident involving the D-21 drone, which remains the highest (and probably the fastest) recorded ejection.

In 1983, a Martin-Baker seat saved its five thousandth life, though this seems to rest on the assumption that all incidents would have resulted in death if not for the ejection. In 1991, US Navy pilot Linda Maloney became the first female Martin-Baker ejectee, escaping from a Grumman EA-6A. (Martin-Baker decided to award her a pewter pin rather than the traditional necktie.) Then, in 1995, the Russian Kamov Ka-50 became the first production helicopter with an ejection seat to enter service. Its rotors were equipped with explosive bolts to jettison the blades before the seat is fired, and pilots of the Ka-50 received the first of many thousands of jokes about helicopter ejection seats. By 2002, they can't be bothered to explain how the seats work.

ABOVE: A British pilot crash-landed in a bomb-laden Harrier GR.9 at Kandahar, Afghanistan, in 2009. After staying with his plane to steer it away from civilian aircraft, he ejected in his Martin-Baker seat as flames began to engulf the cockpit.

TOP 10
INCREDIBLE CANCELLED MILITARY AIRCRAFT

For every military aircraft that makes it into service, a thousand projects live on only as tattered blueprints in filing cabinets, gung-ho artworks or lonely prototypes sitting silently in museums. To be sentimental about killing machines that never were may seem perverse, but the following ten aircraft inspire tantalising speculation of what could have been. Here we take a ghoulish stroll through the graveyard of cancelled military aircraft. (You'll be delighted to hear we haven't included the Avro Arrow, BAC TSR-2 or Northrop F-20.)

'In a war, airbases would be priority targets and, however good a fighter was normally, it would be utterly impotent if it had no runway to take off from'

10. Sud-Est S.E.5000 Baroudeur (1953)
'Jet dirtbike'

Aeroplane designers hate wheels. Wheels are for cars. The weight and complexity of a retractable undercarriage is a huge nuisance. Why not do away with them altogether? The wartime Germans were very keen on this idea and built a series of aeroplanes that took off from trolleys. The aircraft would simply uncouple itself from the trolley as it took off, the trolley remaining behind on the runway, then land on simple skids. A trolley with big chunky tyres frees an aeroplane from the need for vast, vulnerable runways. In a war, airbases would be priority targets and, however good a fighter was normally, it would be utterly impotent if it had no runway to take off from. Mindful of this problem, and fearing the technological hurdles of vertical take-off and landing, Sud-Est turned back to the 'trolley dolly' concept to create the Sud-Est S.E.5000 Baroudeur ('Adventurer'). The aircraft took its first flight on 1 August 1953. It was superb: trolley take-offs proved effortless, skid landings a delight (even in crosswinds). It could be rapidly rearmed and refuelled, and would have made a superb tactical fighter. If required, the trolley could even be rocket-assisted.

It wasn't perfect, however, explains aviation journalist Jon Lake. 'The Baroudeur would have still needed a very long, very level, very smooth runway – just not a concrete one. It also needed heavy lifting gear to get the aircraft back onto its trolley. And it couldn't taxi under its own power or risk landing on a conventional airfield for fear of ripping up the concrete.'

This 'jet dirtbike' never made it into service, usurped by a generation of concrete-loving fast-jets.

OPPOSITE: No aircraft was ever more impressive than the astonishingly six-engined trisonic XB-70 Valkyrie bomber. It would, however, be far too easy to place this North American marvel in the top slot.

ABOVE: Skids and drogue parachute deployed, the Barodeur prepares to land. Designed to land almost anywhere, it once took off from La Baule beach in western France, barely escaping the incoming tide.

BELOW: Maintaining squadrons of eight-engined fighters such as the XF-109 would have been a nightmare for ground crew.

9. Bell XF-109 (never flown)
'Bell époque'

In 1955, the USA's Navy and Air Force approached the Bell Aircraft Corporation with a far-out idea: design a Mach 2 fighter capable of launching and landing vertically. Bell dutifully drew up a design for what it unofficially called the XF-109.

At 59 feet (18 m) long, the XF-109 featured a startling eight J85 turbojets – four afterburning motors arranged two apiece in rotating wingtip nacelles, plus another two afterburners in the rear fuselage and a pair of non-afterburning J85s pointing downward behind the cockpit. The XF-109 was clearly ahead of its time. The US Navy and Air Force both lost interest and the military cancelled the Bell jump-jet in 1961 before the company could even build any prototypes. The Harrier, the world's first operational vertical-take-off-and-landing fighter, flew for the first time in 1967. The subsonic Harrier was far less ambitious than the XF-109, perhaps demonstrating the limit of what was practically possible using the technology of the time.

8. Douglas XB-42 Mixmaster (1944)

'Mixmaster Flash'

The remarkable XB-42 was in many ways the most advanced piston-engined warplane ever flown (though the Republic XF-12 Rainbow might be a rival for this title). As the historian René J. Francillon puts it in his book *McDonnell Douglas Aircraft Since 1920*, 'the XB-42 was as fast as the Mosquito B.XVI but carried twice the maximum bomb-load . . . Furthermore, the Mixmaster had a defensive armament of four 0.50-inch machine guns in two remotely-controlled turrets whereas the Mosquito B.XVI was unarmed.' A variety of offensive gun options were considered, including sixteen .50-cals or two 37-mm cannon. Meanwhile, the updated XB-42A prototype, with its improved engines and Westinghouse 19XB-2A turbojets, had a top speed of 488 mph (785 km/h) and a maximum range of 4,750 miles (7,644 km). The Mixmaster was superb, but at the risk of being too positive, it should be noted that it suffered terrible adverse yaw. Also, while the aforementioned Mosquito could operate from grass airfields, the XB-42 needed miles of flat smooth concrete due to its limited ground clearance.

By the time the war ended, the US Army Air Force could afford to wait for the inevitable arrival of the jet bomber, but the Mixmaster offers a tantalising insight into how military aircraft may have evolved if the piston age had lasted a little longer.

ABOVE: The superb XB-42 had everything on its side other than timing. By 1944, the fastest aeroplanes were jet- or rocket-powered, and however clever and aerodynamically slick designers could make piston-engined aircraft, they were not the future. The additional jets of the XB-42A (lower photo) were not enough.

RIGHT: The story of a fantastically futuristic design losing out to a more conservative rival would be seen again in the YF-22/YF-23 ATF competition of the early 1990s.

7. Martin XB-51 (1949)

'No cigar for the silver cigar'

Blessed with one of the most exotic configurations from a wildly imaginative crop of experimental bombers from various manufacturers, the XB-51 was, frankly, a bit of a dud. Originally designed as a low-level bombing and close-support aircraft, it wound up being considered instead in a 1950 US Air Force competition to find a replacement for the B-26 Invader as a night tactical bomber, alongside the North American B-45 and AJ-1 Savage, and the English Electric Canberra.

The XB-51 featured an engine installation unlike any other, with two General Electric J47 engines mounted in a 'chin' position on the fuselage sides, and a third located in the rear fuselage and fed by a dorsal intake. This arrangement enabled a very clean and thin swept wing to be mounted on the fuselage in a mid-wing position. The wing featured a large-span slotted flap, full-span leading edge slats and variable incidence.

The Air Force's requirements were a service ceiling of 40,000 feet (122 m), a maximum speed of 550 knots (1,019 km/h) and a range of about 1,000 nautical miles (1,852 km), together with all-weather and night operation from basic airfields. Against this specification, to quote *Post-World War II Bombers* by Marcelle Knaack, 'The B-45 was too heavy, and the AJ-1 was too slow.' The competition came down to a fly-off against the Canberra, which had created a sensation by flying non-stop and unrefuelled to the US from Europe – the first jet aircraft to do this.

In the fly-off, the XB-51 lost out to the Canberra, which could exceed the ceiling required by 15,000 feet (4,572 m) and offered double the required range. Although slightly faster at low level, the relatively high wing-loading and low fuel capacity of the XB-51 meant it was beaten by the Canberra on range, ceiling and payload, despite appearing a far more futuristic design.

'The US Air Force wanted a turbojet-powered heavy strategic bomber to lug atomic bombs across oceans'

6. Convair YB-60 (1952)
'Dr Estrangedlove'

In the early 1950s, the US Air Force wanted a turbojet-powered heavy strategic bomber to lug atomic bombs across oceans. Convair had built the piston-engine B-36 aircraft for the Air Force and decided that simply swapping out the B-36's prop motors for jets (among other modest change) would suffice to produce a new bomber.

The result was the YB-60, a 171-foot (52 m) monster of a warplane sporting eight J57 turbojets, which enabled it to fly 2,900 miles (4,667 km) at a cruising speed of 467 mph (752 km/h) while lugging 36 tons (36,577 kg) of bombs. Impressive, but not as impressive as the performance of the YB-60's most direct competitor, Boeing's B-52: the eight-engine B-52 cruises at 525 mph (845 km/h) over a distance of 4,500 miles (7,242 km) while carrying 35 tons (35,561 kg) of bombs.

The first of two YB-60 prototypes took off on its inaugural flight in April 1952, but the US Air Force cancelled the YB-60's test programme in January 1953. B-52s, however, remain in the US inventory.

TOP: The YB-60 was not a bad aircraft, but it was in direct competition with what would prove one of the most durable aircraft designs in history, the B-52. The silver lining for the YB-60 was that the design was spared the horror of the US's disgraceful actions over Southeast Asia in the 1960s and early '70s.

ABOVE: Northrop proposed the Fang in response to an early 1953 USAF request for a high-altitude day fighter, and though the vacancy was eventually filled by the F-104 Starfighter, this design is a fascinating insight into Northrop's thinking.

5. Northrop N-102 Fang (never flown)
'The Bloody Fang'

German-born Ed Schmued designed the North American P-51 Mustang fighter that helped the Allies win the Second World War. A decade later, in 1952 and now working for Northrop, Schmued outlined a simple, single-engine jet fighter: the N-102 Fang.

Forty-one feet long, powered by a single J79 turbojet and sporting a simple delta wing, the Fang bucked the trend towards bigger, heavier and more complex fighters. Northrop built a mock-up and pitched the N-102 to the US Air Force in 1953 and to the Navy in 1954. Ultimately, both branches opted for bigger fighters such as the F-4, which was roughly twice the Fang's size and boasted two J79s.

But Northrop didn't entirely give up on the idea of a small, simple fighter. The company's F-5 family of fighters, including the F-5A, the much-improved F-5E and variants, and even the prototype F-20, all owe their design philosophy to Schmued's N-102. F-5s remain in service today, unless you've found this book in a charity shop many years in the future.

3. Lockheed CL-1200 Lancer (never flown)
'Cosmic lancer'

In the late 1960s, Lockheed saw an opportunity. Anticipating worldwide demand for 7,500 advanced but – in the company's own words – 'reasonably-priced' jet fighters over the next decade, in 1971 it began circulating a proposal for an improved, safer derivative of the speedy but notoriously hard-to-fly F-104 Starfighter.

Lockheed's Skunk Works division, with famed designer Kelly Johnson at the head, enlarged the F-104's wing and fin, shifted the tailplane lower in the fuselage, tweaked the engine inlet, added internal fuel capacity and replaced the F-104's J79 engine with a TF33. The result, the CL-1200 Lancer, was in theory more manoeuvrable and controllable than the F-104 and cost around $2 million per copy, assuming a large production run, compared to $2.4 million for a new F-4E at the time. With 46 per cent more fuel on board than the Starfighter, half its required runway length and an even faster top speed, of Mach 2.5, it would have been a very impressive machine.

Lockheed entered the CL-1200 into the US military's International Fighter Aircraft competition, which aimed to select an export warplane for America's allies. But Northrop's F-5E won the contest and Lockheed scrapped the CL-1200 concept, having only ever produced a mock-up of the plane.

4. Convair Model 49 (never flown)
'Ring-wing transformer'

In the 1960s the US Army were growing sick of their dependence on inappropriate US Air Force aircraft for close air support. Aircraft like the Republic F-105 Thunderchief were simply too fast and too vulnerable to support troops on the ground effectively. Instead, the US Army wanted the versatility and forward-basing possibilities of a vertical take-off platform with the ability to hover. To excel in its tough, close-support role, the type would need to be heavily armed and armoured. This need was expressed formally as the Advanced Aerial Fire Support System, or AAFSS.

Convair, a company famed for its adventurous designs, responded to the Army's AAFSS requirement with typical ambition. Drawing on their experience with the tail-sitting XFY-1 Pogo, they proposed a two-man 'ring' (or annular) wing-ducted-fan design quite unlike anything else in service, though somewhat similar to the experimental SNECMA C.450 Coléoptère. The concept was bizarre in appearance but Convair believed it was the perfect configuration for an aircraft combining a helicopter's unusual abilities with some of the offensive features of a military ground vehicle. One of the greatest challenges was to create a cockpit that tilted, so the pilot was not facing the sky during the take-offs and landings. This necessitated a complex hinged forward-fuselage, giving the type its distinctly Transformer-like looks.

ABOVE: Quite how survivable the Model 49 would actually have been on a 1970s battlefield is questionable. From appearance it looks distinctly vulnerable to asymmetrical attacks from Ewoks.

BELOW: The CL-1200 was an ingeniously economical upcycle of the F-104 by Kelly Johnson. The reinforced windshield from the F-104S was to be included to withstand the heat of even higher-speed flight.

2. Dassault Mirage 4000 (1979)
'Hyper-electric cake slice'

France's Mirage 2000 has been described by many fighter pilots as the perfect flying machine. Its ferociously high performance and almost telekinetic responsiveness have left pilots of all nationalities giddy with love and respect for the 'electric cake slice'. So imagine a 2000 with twice the power and you have a pretty spectacular aeroplane: the 4000, which first flew in 1979, in the same heavyweight class as the McDonnell Douglas F-15 Eagle and Sukhoi Su-27. The Mirage 4000 was one of the first aircraft to incorporate carbon-fibre composites (to keep weight down) and probably the first to feature a fin made of this advanced material.

Thanks to its light structure and powerful engines, it had a thrust-to-weight ratio that exceeded 1:1 in an air-to-air load-out.

On its sixth test-flight it reached 50,000 feet (15,240 m) at Mach 2 in 3 minutes 50 seconds. The 4000 was agile, long-ranged and able to haul an impressive arsenal. Its capacious nose could have held an advanced long-range radar. But the French air force didn't want it, Iran – another potential customer – had a revolution, and Saudi Arabia – also on the lookout for a heavy fighter – opted instead for the F-15. Despite its obvious potential, the Mirage 4000 failed to find a customer, which was an enormous kick in the nuts for Dassault, as the company had largely privately funded the type's development.

1. Lockheed RS-12 (never flown) and F-12 (1963)
'Kelly's killer'

In January 1961, Lockheed's legendary aeroplane designer Kelly Johnson delivered an unsolicited proposal to the US Air Force. His idea was to take the Mach 3 A-12 spy plane – the predecessor of the iconic SR-71 Blackbird, which Kelly had designed for the CIA – and modify it to become a very fast strategic bomber, designated RB-12. More or less in parallel, Johnson was working on a missile-armed fighter version of the A-12, which would have been designated F-12 had it entered service.

The Air Force liked Johnson's original bomber idea, but counterproposed with a slightly altered design that it called the RS-12: take the A-12's sled-like titanium airframe with its powerful J58

turbojets and add a sophisticated, long-range radar and a nuclear-tipped air-to-ground missile based on the AIM-47 (originally known as GAR-9) air-to-air missile that also armed the F-12.

The plan was for the RS-12 to penetrate Soviet air space at Mach 3.2 and 80,000 feet (24,384 m), and fire a single missile from 50 miles (80 km) away, striking within 50 feet (15 m) of its aim-point within a Soviet city. The Department of Defense ultimately cancelled the F-12 on cost grounds and opted not to proceed with the RS-12, as ballistic missiles were beginning to supplant manned bombers. The Air Force did ultimately acquire the SR-71 reconnaissance version of the A-12, however, and operated it into the 1990s.

ABOVE: An imagined Mirage 4000 of the Escadron de Chasse 2/5 Île-de-France deployed to Saudi Arabia to enforce the 'No Fly Zone' in 1992 with an air-to-air load-out of Magic and Super 530 missiles. (As an aside: in 2019 Italian police arrested some far-right terrorists for possession of an ex-Qatari Super 530 missile.) Markings show this aircraft has already performed air-to-ground missions, revealing a multi-role tasking.

TOP 10
CANCELLED BRITISH FIGHTERS

From the dawn of the aeroplane until the 1960s, Britain produced world-class fighter aircraft. As well as the designs that actually felt the air beneath their wings, there is a tantalising treasury of designs that never made it. Here are ten of them.

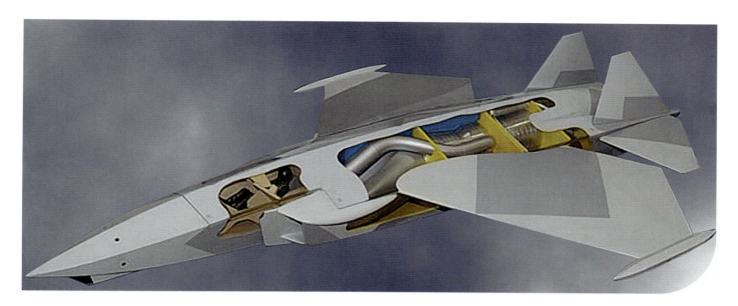

10. British Aerospace P.125 (never flown)

'Have not glass'

The long history of British expertise in stealth technology has not been discussed a great deal. Britain pioneered radar-absorbent material for aircraft, worked on reduced radar-observability for nuclear warheads in the early 1960s and was able to create a world-class stealth test-bed in the Replica model. Prior to Replica, in the 1980s Britain was working on an aircraft concept so advanced it was classified until 2006: the BAe P.125.

The P.125 study was for a stealthy supersonic attack aircraft to replace the Tornado. It was to be available in both a short take-off and vertical landing (STOVL) version and a conventional variant. The conventional variant would feature a central vectoring nozzle, the STOVL version would have three vectoring nozzles. In some ways the P.125 was more ambitious than the F-35; the aircraft was to have no pilot transparencies, with the reclined pilot instead immersed in synthetic displays of the outside word.

The absence of a cockpit transparency on the P.125 was probably intended to protect the pilot from laser-dazzle weapons (a weapon inaccurately feared to be in widespread use by the Soviet Union). Even now, a synthetic worldview cockpit is considered a daunting technological prospect; why BAe didn't opt for an unmanned configuration remains something of a mystery.

It is likely that this formidable interdictor would have been even less visible to radar than the F-35 (though the absence of planform alignment is noteworthy). Despite its 1980s vintage,

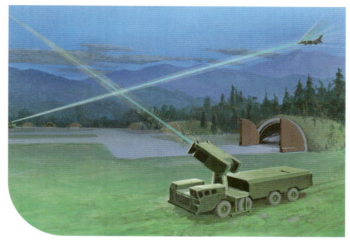

'The absence of a cockpit transparency on the P.125 was probably intended to protect the pilot from laser-dazzle weapons (a weapon inaccurately feared to be. in widespread use by the Soviet Union)'

many of its low-observable features are reminiscent of today's latest fighters, while other features, such as its unorthodox wing design, are unique. The project was quietly dropped when Britain joined the Joint Strike Force programme in the 1990s.

9. British Aerospace P.1214-3 (never flown)
'Garryhawk'

The P.1214 studies tried to solve the inherent limitations of the Harrier concept. The Harrier's Pegasus engine, with its steerable thrust, blesses the Harrier with the ability to take off and land vertically – and even fly backwards. Unfortunately, you can't put conventional afterburners on a Pegasus engine. There are several reasons for this: the hot and cold air is separated, the inlets do not slow the airflow sufficiently for serious supersonic flight, and the jet-pipes would be too short. Conventional afterburners on a Harrier would also set fire to everything (it was tried from the 1960s and proved problematic). This is a shame as a Harrier desperately needs thrust on take-off and could do with the ability to perform a decent high-speed dash.

Though conventional afterburners are out of the question, you could, however, use plenum chamber burning (PCB). This technology was developed for the Mach 2 Hawker Siddeley P.1154 (think lovechild of a Harrier and a F-4, with the wingspan of a Messerschmitt Bf 109), which never entered service. PCB chucks additional fuel only into a turbofan's cold bypass air and

ignites it (a conventional afterburner puts the burning fuel into the combined cold and hot gas-flows). This is great, but how do you incorporate this into swivelling nozzles without destroying the rear fuselage with heat and vibration? BAe thought it had found the answer: get rid of the rear fuselage altogether and mount the tail onto two booms. Worried that this already eccentric idea might seem too conventional, BAe decided to add an 'X-wing' configuration with swept-forward wings (which were in vogue in the early 1980s). This did produce the coolest fighter concept of the 1980s, even in the -3 variant shown in the image above, which had conventional tails.

There's another fundamental issue around intake design, too. The Harrier has a short duct and auxiliary suck-in doors to get maximum mass-flow at zero speed. At high speeds, the intake spills flow outwards around the lip, causing a huge amount of drag. Matching intake performance at either end of the speed range is an issue for any direct-thrust STOVL aircraft. The X-32 tried to solve this with a cowl that translated forwards at low speed; the F-35 solved the problem by having a separate lift fan. On the P.1214, the inlet was longer than on the Harrier to manage inlet distortion better than with the short duct, but it would still have spilled at a higher Mach.

The P.1214 would have been extremely agile (and short-ranged), probably much like the Yak-41. The P.1214 lost its swept-forward wings when further studies revealed them to be of no great value. It now became the P.1216, which was intended to satisfy the US Marine Corps and the Royal Navy's desire for a supersonic jump-jet (a need eventually met by the F-35B). A full-sized wooden P.1216 was built to distract Thatcher from stealing children's milk; predictably (as it was British), the whole project was scrapped. This was arguably a good thing, as British military hardware testing and development was at its lowest ebb in the 1980s (see the Nimrod AEW3, SA80 rifle family, Foxhunter radar, ADEN 25 cannon and Harrier GR.5 compared to the US AV-8B, etc. for details).

Supersonic aircraft have their jet exhausts at the back, and there is a simple reason for this: anything in the way of the jet efflux will be exposed to a destructive barrage of heat and vibration. This presents a problem to supersonic STOVL designs wishing to use vectored thrust – to have sufficient thrust and acceleration, jet-flow far hotter than the Harrier's is required. One way to solve this is to have no rear fuselage.

We asked aerodynamicist and aircraft designer Stephen McParlin if the P1214-3 would have been any good and if it would have worked in reality. He replied: 'Yes, it would. . . and so would the forward-swept wing. Kingston had a very experienced and capable team, but the MoD were looking to share costs and risks with overseas partners, and low observables were bubbling under. The subsequent looks at Sea Harrier Replacement were driven by affordability, so new wings on a minimum-change Harrier fuselage was the limited scope.' And so it was that this extremely promising idea was lost to history. In response to our unofficial contest to name this aircraft for the book, Stephen said: 'I'd like to call it Garry, after Garry Lockley, who was the aerodynamics lead on this and P.1216. He was a lovely bloke and good friend, who loved real ale and cricket, and he died way too young, of early-onset Parkinson's.'

ABOVE: By landing on its hull rather than large floats, the SR.A/1 avoided the performance penalties of a draggy design. It also meant that the aircraft was a flying boat seaplane as opposed to a floatplane.

8. Saunders-Roe SR.A/1 (1947)
'The Squirt Queen'

The aircraft was first proposed in mid-1943, the combination of jet-engine speed and the flexible basing options of a flying boat being regarded as advantageous in the Pacific theatre. Development lagged, however, and the aircraft didn't fly until 16 July 1947. Three aircraft were built, two of which crashed. The simultaneous development of the Princess by Saunders-Roe contributed to the slow work on the SR.A/1, and this was compounded by the decision by Metropolitan-Vickers to cease turbojet engine production.

Although exhibiting quite sprightly performance, by the time the SR.A/1 had flown, the Pacific war was over and there was no more need for such an aircraft. In addition, the Fleet Air Arm was operating numerous aircraft carriers, and the development of capable jet-powered, carrier-based aircraft allowed power projection without the need for airfield construction. Additionally, of course, the large number of airfields constructed during the war also provided many basing opportunities for conventional land aircraft.

7. Saunders-Roe SR.53 (1957)
'Death Roe'

Fast, but outpaced by changes in the threat – and in government policy – the Saunders-Roe SR.53 was proposed to meet a requirement for a point-defence interceptor capable of climbing to 60,000 feet (18,288 m) in 2 minutes and 30 seconds. The driver for this need was concern about the threat posed by Soviet bombers armed with nuclear weapons.

The SR.53 was a compact, delta-winged, mixed-power aircraft with a 1,640 lb thrust Rolls-Royce Viper jet engine and a 8,000 lb de Havilland Spectre rocket. The armament was intended to be the Blue Jay infrared air-to-air missile. The operational concept was to climb to altitude using the rocket motor, accelerate up to a maximum speed of Mach 2.2, complete a ground-guided interception and then return to base using the jet engine.

The contract to develop the aircraft was signed on 8 May 1953. Although Saunders-Roe's initial schedule called for a first flight in July 1954, development of the aircraft and its rocket motor took longer than expected, and first flight did not occur until 16 May 1957, with a second prototype following in December of the same year. The aircraft was reported as being pleasant and easy to fly. The second prototype was lost in a fatal, aborted-take-off accident in June 1958, and the programme was eventually cancelled in July 1960, after fifty-six test flights. The highest speed reached in the flight-test programme was Mach 1.33 (some sources put this even faster at Mach 1.45).

During the seven-year development and flight programme, a great deal of change had occurred in aerospace capabilities: jet-engine development had produced far more powerful and reliable engines; radar had improved its ability to detect targets at long range; the Soviets had moved towards the development of standoff weapons; and surface-based guided missiles had improved in capability. These technical advances effectively invalidated the operational concept for the SR.53. In future, it would be possible, and necessary, to defeat threats at a greater distance before the release of nuclear standoff weapons, and there was no way a short-range point-defence interceptor such as the SR.53 could achieve this.

Furthermore, the first flight of the aircraft occurred just two months after Duncan Sandys' 1957 Defence White Paper, which suggested that new manned aircraft were no longer required for air defence and that surface-based air-to-air missiles would in future fill this role. The first flight of the SR.53, just after this policy announcement, could not have been more badly timed, but the basis for the aircraft had already been superseded. The programme left no direct legacy.

It is worth noting that the SR.53's designer, Maurice Brennan, had recognised the need for more range and proper radar long before the MoD did, and had suggested a solution in the form of the SR.177 design. But SR.177, which was rather more refined, was also cancelled in the 1957 White Paper. Brennan had already suffered the heartbreak of having his gargantuan SR.45 Princess axed and the embarrassment of his A.36 Lerwick entering service.

'The first flight of the SR.53, just after this policy announcement, could not have been more badly timed, but the basis for the aircraft had already been superseded'

BELOW: This second prototype SR.53 crashed in 1958, killing test pilot Squadron Leader John Booth DFC. The crash was not the result of the novel propulsion system but a blind flying panel that dropped down behind the control column, and the subsequent failure of the brake chute. The total flying time of this airframe was 5 hours and 15 minutes.

6. Martin-Baker MB3 (1942)
'Valentine's dayfighter massacre'

The MB3 appeared in 1942, the result of a prudent Air Ministry decision in 1939 to obtain a powerfully armed fighter as an alternative to the Hawker Typhoon in the event that the Typhoon programme ran into insurmountable technical difficulties or serious delays. The aircraft that emerged looked sensational, especially when the unprecedented armament of six 20-mm cannon was fitted. Despite looking insane, it was unusually sensible: a multitude of access panels made it far easier to maintain than its contemporaries, and its tough structure (a more advanced version of the load-bearing tubular-box type favoured by Hawker) would have given it greater survivability. It was apparently easy to handle and extremely fast.

Unfortunately, we don't know exactly how fast, because less than two weeks after the first flight the Napier Sabre engine that powered it did what Napier Sabres were doing in droves in 1942 and packed up. The MB3 was destroyed in the subsequent forced landing, which also killed test pilot Valentine Baker (the 'Baker' of Martin-Baker). This was a serious blow to the company and affected designer James Martin (the 'Martin' of Martin-Baker) so much that he devoted the rest of his career to making aircraft safer by developing ejection seats, which Martin-Baker continue to produce to this day.

5. Fairey Delta 3 (never flown)
'The delta belter'

The Fairey Delta 2 experimental aircraft was the first aeroplane to exceed 1,000 mph (1,600 km/h) and in 1956 it took the World Airspeed Record to 1,132 mph (1,822 km/h). It was a beautifully simple design with the delta wing's inherent advantages of low supersonic drag and great structural strength. A year earlier, the Air Ministry had issued Operation Requirement F155T, specifying a supersonic interceptor able to intercept Mach 1.3 bombers at 60,000 feet (18,300 m). After initially proposing a modestly updated weaponised Delta 2, Fairey came forward with the mighty Delta 3 – a vast, super-high-performance interceptor with state-of-the-art technology – and won the contest.

Mixed propulsion, of jet engine and rocket, was necessary to meet the extremely demanding specification, which called for the fighter to reach 60,000 feet at a range of 70 nautical miles (130 km) from base in 6 minutes, at a speed of at least Mach 2. The maximum climb-rate would have been phenomenal, leaving even the English Electric Lightning for dust and even rivalling today's fastest climber, the Typhoon. The thrust levels were astonishing – according to some sources, it was to have two Rolls-Royce RB.122 engines, each boasting a dry thrust of 19,500 lb ft, and 27,800 lb ft with re-heat, greater than the present-day Flanker. And that's not taking into account the additional rocket engines! Not bad for an aircraft that had normal operating weight of just over 50,000 lb (22,700 kg). Its top speed was estimated at between Mach 2.3 and 2.5.

To soak up the heat generated by high-supersonic flight, the centre fuselage, fin and wing were to be built from stainless steel (steel was used on the Bristol 188 and MiG-25 for the same reason), while the parts of the aircraft system requiring the greatest heat tolerance were to be made of titanium alloy. The aircraft was to be armed with two of the giant Red Dean missiles, a unit which, coming thirty years before the AMRAAM and even ten before the AIM-54 missile, was ambitiously planned as an active-radar-guided missile.

Heavy, ultra-high-performance interceptors did not prove popular in the West, however. The North American XF-108 Rapier, CF-105 Arrow, Lockheed YF-12 and Mirage 4000 were all cancelled; they were too expensive and air forces instead opted for more modest interceptors backed up by surface-to-air missiles. The Fairey 3 may have suffered the same fate, had it survived Duncan Sandys' ill-conceived crusade against manned aircraft of 1957, which it did not.

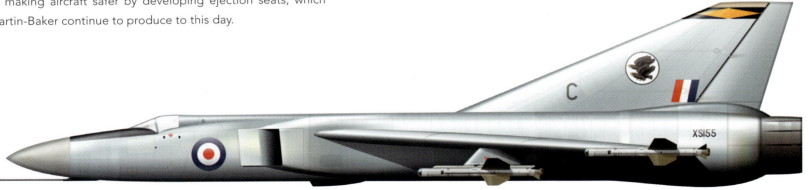

4. Hawker P.1103/P.1121 (never flown)
'Super Hunter'

Hawker tried to turn their highly successful transonic fighter, the Hunter, into a supersonic aircraft as their offering in response to the Air Ministry's Operational Requirement F.155. The limitations of the Hunter – the lack of air-to-air missile capability, decent enough radar and the ability to reach supersonic speeds – would be addressed by a radical redesign. The new fighter interceptor, the P.1103, would include a completely new fuselage and wing (changing more profoundly than Trigger's broom in *Only Fools and Horses*), a seat for the radar operator, a far more powerful engine and missile armament. To make room for the new radar, a chin intake was adopted. As with the Vickers Type 559, the design included booster rockets for added climb-speed, though in practice operational versions would have likely omitted them.

The P.1103 was quickly knocked out of the Ministry's contest, partly due to the Ministry's contention that Hawker had not embraced, nor even fully understood, the idea of the aircraft as a 'weapon system'. But Hawker had faith in the design and continued with it as the self-financed P.1121. Power was to come from a single de Havilland Gyron jet engine (as this is Hawker we're talking about and their fighters were always single-engined) and the aircraft was to be armed with Red Top missiles, rockets and ADEN 30-mm cannon. Maximum speed was estimated at an astonishing Mach 1.35 at sea level – and a rather more believable Mach 2.35 at higher altitudes. But the Air Staff only reluctantly reconsidered the design before again turning their noses up at it. The 1957 Defence White Paper put further nails in its coffin, though Hawker persisted with the idea for another year before finally giving up.

The design would probably have inherited some of the fine handling characteristics Hawker had instilled in earlier aircraft, such as the Hunter and Fury. The somewhat generous wing area and decent thrust-to-weight ratio (for the time) meant the 'Super Hunter' should have enjoyed good turn rates for its generation. A well-balanced, sensible design with impressive performance, the P.1121 could have enjoyed good export sales and offered the RAF a more versatile and combat-effective fighter than the Lightning, and one that would have likely proved formidable in both the air superiority and ground-attack roles.

BELOW: An RAF 34 Squadron P.1121 armed with WE.177 nuclear bombs from the 1960s. The type's great speed (topping out at a predicted Mach 2.5) would have proved useful in its intended roles, as both interceptor and nuclear-strike aircraft.

'The P.1121 could have enjoyed good export sales and offered the RAF a more versatile and combat-effective fighter than the Lightning'

3. Martin-Baker MB5 (1942)
'Martin-Baker Tie Fighter'

Despite the aforementioned crash of the MB3 in 1942 due to the failure of its Napier Sabre engine, it was apparent that the plane was worthy of further development. Martin-Baker proposed a Rolls-Royce Griffon-powered version, the MB4, but a more thorough redesign was favoured by the Air Ministry and the MB5 was the result. A fair contender for the best British piston-engined fighter ever flown, the MB5 was well armed (though with the less-impressive total of four, rather than six, cannon), very fast, and as easy to maintain as its predecessor. Flight trials proved it to be truly exceptional, with a top speed of 460 mph (740 km/h), brisk acceleration and docile handling. Its cockpit layout set a gold standard that Boscombe Down recommended should be followed by all piston-engined fighters.

The only thing the MB5 lacked was good timing. It first flew two weeks before the Allied invasion of Normandy. Appearing at the birth of the jet age, with readily available Spitfires and Tempests, both of which were themselves excellent fighters, there was never a particularly compelling case for producing the slightly better MB5. There is also a suggestion that the MB5 never received a production order because on the occasion it was being demonstrated to assorted dignitaries, including Winston Churchill, the engine failed. If this is true, it must rank as the most pathetic reason for non-procurement of an outstanding aircraft in aviation history.

Other contenders for the title of 'Best British Piston-Engined Fighter' would include the Spitfire Mks.18 and 24, the Hawker Sea Fury and the de Havilland Hornet, though the definitive answer can only be decided in a six-pint-long pub debate. The MB5 must be included in this debate as it boasted the following: better range and armament than any Spitfire mark; slightly higher top speed, better range and better ceiling than a Sea Fury; better manoeuvrability than a Hornet, and it was also cheaper and flew earlier.

ABOVE: James 'Jimmy' Martin stands by his masterpiece, the formidable MB5.

1. Hawker Siddeley P.1154 (never flown)
'The Hyper Harrier'

By the mid-1950s, it was obvious to many Western military planners that, in the event of war, Warsaw Pact forces would quickly obliterate NATO airbases. For NATO aircraft to mount counter-attacks (some with tactical nuclear weapons), they would need to operate from rough, unprepared airstrips. This capability could turn air arms into survivable 'guerrilla' forces able to fight on after the apocalypse. Vertical take-off and landing (VTOL) was also tempting to many navies, as it could eliminate the traditional hazards of carrier landing. If an aircraft could stop moving forward before it landed, the task of settling on a tiny, pitching deck would be far easier. Likewise, it could liberate ships from the need to carry enormously heavy catapult-launch systems and could even allow small ships to carry their own high-performance escort aircraft.

The prospect of providing NATO with a common fighter soon attracted most major Western aircraft companies. NATO's Basic Military Requirement 3, the third in a series of documents drawn up by the alliance to detail the specifications of future combat aircraft designs, became the biggest international design competition ever held. Two months later, NBMR-3 was split into two: AC 169a would cover a F-104G replacement, and kept the original demands; AC 169b was to be a Fiat G.91 replacement. AC 169b differed from AC 169a in calling for a lower payload-range requirement of 180 nautical miles (333 km) with a store of 1,000 lb (454 kg).

At this point, OR345 was dropped in favour of NBMR-3. Hawker Siddeley's bid was the monstrous P.1154 equipped with the

BELOW: The P.1154 as it would have appeared in RAF (upper) and Royal Navy service. Had it served in the Falklands War of 1982, it would likely have utterly dominated Argentinian aircraft to an even greater degree than that managed by the far inferior Sea Harrier.

insanely powerful Bristol Siddeley BS.100 engine. Hawker's and Bristol's P.1154 was declared the winner, but history repeated itself. Though nobody was tied to buying the winners of NBMR contests, it seems unfair that no country outside Britain was forthcoming in wanting to invest in the aircraft. At least Hawker still had a generous MoD budget to work with, and the type was elected to replace RAF Hunters and the Royal Navy's de Havilland Sea Vixens – what else could go wrong? Two things. The first was the differing needs of the Royal Navy and the RAF. The RAF wanted a single-engined single-seater. The Navy wanted a two-seat, twin-engined aircraft. To some degree, both of the Navy's wants may have been driven by safety regulations regarding nuclear-armed aircraft (though they were content to have the single-seat Scimitar carried the Red Beard tactical nuclear bomb). The Navy was also impressed by the F-4 Phantom II, and there were some within the Admiralty who considered this a safer option. Giving the P.1154 twin engines would involve a complex modification of the design. The BS.100 was too big to be fitted in pairs, so the Rolls-Royce Spey engine was selected. To stop a twin-engined P.1154 flipping over in the event of a single-engine failure, a complicated twin-ducting concept was added. The Royal Navy also wanted a larger radar.

On top of this, P.1154 threatened the existence of the Navy's big carriers: if these new machines could take-off in next to no distance, why did the Navy need massive, expensive carriers at all? It should be noted that the Navy intended to catapult-launch their P.1154s, using an US style of operation – the Navy's self-preservation instinct was kicking in. While the RAF P.1154s could have been made to work (with limitations), many – even at Hawker – doubted the viability of the naval variant.

If the first major problems facing the P.1154 were inter-service differences, the second set were technical. The temperature of the hot, after-burning exhaust that the P.1154 would be firing from its front nozzles down onto runways or carrier decks would be great enough to melt asphalt or distort steel (the Yak-141 would later encounter similar problems). The exhaust would also churn up a potentially dangerous cloud of any present dirt. Added to this was hot gas reingestion (HGR); the aircraft would be 'breathing in' its own hot exhausts on landing. This recirculating hot air would raise the temperature in the engine to more than it liked – another very serious problem.

On 2 February 1965, the incoming Labour government, led by Harold Wilson, cancelled the P.1154 on cost grounds. Was this to be the end of VTOL and STOL fighters? Fortunately not. While the P.1154 was being designed, Hawker had been busy developing the P.1127 into the Kestrel, with the help of funds from Britain, West Germany and the USA (initially from the US Army). This, of course, led to the Harrier, the famous jump-jet which today remains in service with the United States, Spain and Italy.

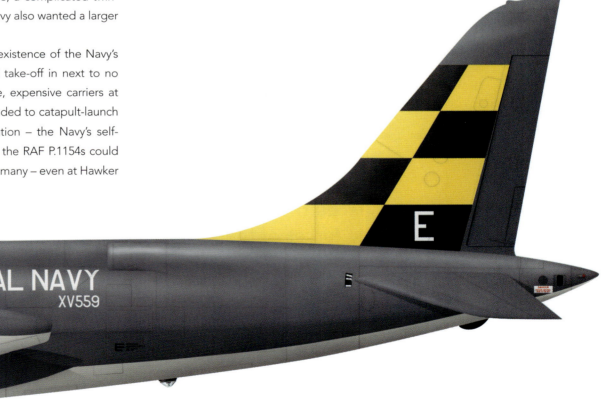

FLYING AND FIGHTING IN THE SU-15

INTERVIEW WITH COLD WAR FLAGON PILOT VALERI SHATROV

Soviet interceptor pilots faced the daunting task of defending the largest nation ever seen against the combined forces of the world's most technologically advanced nations. The awesome responsibility of preventing nuclear annihilation from US Air Force B-52 bombers, countering the impossible nuisance of snooping Mach 3 SR-71s, and air-to-air combat with F-4 and F-15s, were formidable tasks that the skilled pilots of the Soviet Air Defence Forces (PVO) trained for in earnest. In the Cold War, the backbone of their manned air defence was 1,300 Sukhoi Su-15s (NATO aircraft codename: 'Flagon'). Former Soviet Su-15 pilot Valeri Shatrov spoke to Hush-Kit about his experience of flying this charismatic interceptor.

If I had to choose three words to describe the Su-15, I would choose reliability, manoeuvrability and beauty. High reliability was one of its best traits – I do not remember any tech failures. Su-15 crashes were very rare and were mostly caused by the human factor.

The main purpose of the Su-15 was to intercept air targets, by day or by night. The radar sight 'saw' a little worse than the radars of our potential enemies, but the long-range and short-range air-to-air missiles worked well, at least from my experience from training launches. Thirty years or more after flying the Su-15, when I was at one of the air shows in the United States, I had a chance to meet and speak with a SR-71 pilot. We were joking, exchanging impressions about our winged machines. I confessed to him that if I was in my Su-15 attacking him from the front hemisphere, I probably could have taken him out – but there was no chance of intercepting such a fast aircraft in the chase!

I had a comfortable cockpit; everything in it was simple and easy to find. In those times, there were no autopilots and navigation systems as we know them today, so we flew without any. Navigation was done with the aid of often-barely-legible pilots' written notes! Very primitive by modern standards. Again, the lack of modern navigation instruments and landing systems required the pilot to maximise his skill in order to land the plane in minimum weather conditions. And it was nothing – it was OK for us. There were times you *had* to land below minimum weather conditions.

STARTING ON THE SU-15

I got on to the Su-15 immediately after graduating from the Armavir flight school (or the *Armavir Higher Military Aviation Red Banner School for Air Defence Pilots*, to give it its full title). At Armavir I trained on the L-29, MiG-15UTI and MiG-17. After completing a training course that included live-ammo gun-firing on ground targets, night flights and flights in challenging meteorological conditions we graduates were awarded the qualification of 'Military pilot of the Third Class'. Following graduation in 1976, I was appointed to the 302nd Fighter Regiment [302 ИП in Russian] at Perevaslavka 2, located in the far-eastern region of the country, near the city of Khabarovsk (this regiment doesn't exist any more). My first impression of the type was the elegance of its extremely-swept wing and tail. The plane was very beautiful and extremely graceful. The Su-15 was a typical fighter of its time; we climbed into the cockpit using the ladder. The cabin was comfortable and relatively spacious. Even in winter, wearing fur jacket and boots, I didn't feel any discomfort inside the cockpit. Take-off was not difficult, but landing was. Its high approach-speed required precision-landing and the timely release of the braking parachute.

It was a good plane and, as I've mentioned, exceptionally reliable. I always said that I loved it like a woman. And when you turned on the afterburner on take-off, you got a hefty push in the back from the ferocious power of the two engines. The relatively long fuselage and delta wings allowed you to make a large number of barrel rolls with any rotation speed. What I liked most about the Su-15 was its ability to perform complex aerobatics at full power.

Though all Sukhoi aircraft cockpits differ, they all share some familiar traits. They are all quite spacious, and the instrument panel is mounted a considerable distance from the pilot, which allows you to switch attention from one device to another with only a slight angular movement of the eyes.

ABOVE: This Su-15 is in the hands of the Su-15 technical and operational unit. The Soviet method of aircraft evaluation emphasised putting aircraft in frontline use as swiftly as possible.

'Two aircraft were always in a state of quick-readiness alert: one for high-altitude target interception and the second for the low-speed interception of reconnaissance air balloons'

КОНКУРС «ТИХООКЕАНСКОЙ ЗВЕЗДЫ»

ПЕРЕХВАТЧИКИ. Военные лётчики коммунисты Виктор Дригин, Юрий Сухо-носов, Валерий Шатров. Фото Г. Косенкова.

TOP LEFT: A typical day on a Sukhoi Su-15 base: Shatrov posing beside his aircraft in 1978.

TOP RIGHT: A painting by one of the regiment's pilots, now a sculptor and artist, Vladimir Beketov. It is called: *ARKP Su-15-98 while passing the front of the occlusions of a two-centre pressure depression in the evening at Mach 1.9.*

ABOVE: '*Pacific Star Competition*', a photo contest held by the eponymous newspaper in 1978 (see the emblem on the left). The photo, by G. Konsenkov, is captioned 'Interceptors: military communist pilots Victor Drigin, Yuri Suhonosov and Valeri Shatrov.'

THE WORST THING ABOUT THE AIRCRAFT?

The first of the serial Su-15 modifications, the Su-15TM Flagon-F models, were called 'Hound Dog' for their extremely high landing speed and 'Dove of Peace' for having only two air-to-air missiles. I was lucky – I flew and was on duty with four missiles and two cannon pods. And that version had a leading-edge extension on the wing, somewhat reducing the landing speed.

Also, the aircraft was equipped only with a radio compass; there were no other landing systems.

LIFE ON THE BASE

The day-flight shift began at dawn and lasted 6 hours. Night shift lasted 2 hours in the afternoon – at dusk – and 4 hours at night. For service we wore flight suits and added jackets in autumn and winter. We had boots, and a belt much like a Sam Browne. There was even a joke about that uniform: 'When I wear a belt, I get more and more stupid.' [This is a play on words that doesn't really translate directly into English: the word 'portupeya' (army belt) contains 'tupeyu' which means 'stupid'.] We served in air defence, so we rarely wore everyday clothes.

The regiment was constantly on combat-readiness. Two aircraft were always in a state of quick-readiness alert: one for high-altitude target interception (with four air-to-air missiles), and the second for the 'work' purposes of the low-speed interception of reconnaissance air balloons (four air-to-air

missiles; two GSh-23 cannon pods on hardpoints).

On weekends, pilots with families spent time with their children and wives, walked along the only street of the military town, swam in the river or gathered mushrooms in the forest nearby; or, in winter, skied. Single pilots spent weekends at the dances in the Officers' House, where girls from Khabarovsk and nearby villages came. We drank in our companies – or in company – but on Sunday we'd abstain from drinking – known as 'taking a breath' – if there were flights on Monday, because the doctor would have strict controls before each flight.

Any flight was a feast for a military pilot in my time. I can't single out one as being the most interesting mission. I remember being very fond of any training missions that included live-missile launches on target aircraft, combat-duty missions, flights in minimum weather conditions and night flights. I have strong mental images of penetrating clouds as a pair. Any flights for a military pilot are a holiday! In addition to training for launches of air-to-air missiles, one had to shoot cannon at ground targets. We had no misses – I guess that says a lot. For its time, the aircraft was well armed and equipped with decent avionics.

As I have mentioned, the aircraft was very manoeuvrable. With the engines set at maximum thrust, the Su-15 allowed you to perform a turn with a roll of almost 90 degrees. As for the service ceiling of the aircraft, I personally gained a height of 23,000 metres [75,459 feet], where 'a little higher: space begins', as we joked in those days. The instantaneous turn-rate was very good, with a G-force of up to 6. The high alpha of the aircraft [its ability to raise its nose relative to the direction of flight] was limited to this maximum overload, but the thrust of the engines allowed any horizontal aerobatics figures without loss

of speed. 'Split S' manoeuvres were allowed from a minimum altitude of 2.5 km up to the practical service ceiling.

There were many simulated interceptions according to the course of combat-training requirements: air combats one-on-one, pair-vs-pair and flight-vs-flight. The 'enemy aircraft' for training were usually Su-15s from our regiment. There were also training flights to practise interceptions against bombers. Usually the target bomber was the Tu-16. For actual missions, the SR-71 was the hardest to intercept.

Actually, I had no chance meetings with the SR-71. They 'teased' our pilots on combat duty, especially in Primorsky Krai [an area in the Russian Far East bordering North Korea]. Sometimes duty vehicles had real interception flights – aircraft were frequently scrambled. The SR-71s were performing flights in a neutral zone but with each launch of our interception aircraft they turned toward their base. In my case, bases in Japan.

In wartime, the Su-15 would have intercepted B-52s, which raises the question of whether the B-52s would be accompanied by a number of escort fighters, which would have drawn us into a missile air battle. While F-4s and Su-15s had the same manoeuvring characteristics and armament, the F-15 would have proved a far more challenging battle opponent. The F-15 was more manoeuvrable than our aircraft, and its electronic equipment was much more modern.

TOP LEFT: Seeing the Su-15 sent to the dustbin of history (Khurba, 1984).

TOP RIGHT: And your humble servant.

LONELINESS AT MACH 3

INTERVIEW WITH MIKOYAN-GUREVICH MIG-25 FOXBAT PILOT AIR MARSHAL SUMIT MUKERJI

High-flying, insanely fast and untouchable, the MiG-25R Foxbat served the Indian Air Force with aplomb. Air Marshal Sumit Mukerji spoke to Hush-Kit about flying the world's fastest operational aircraft.

What were your first impressions of flying the MiG-25R?

A 20-ton aircraft that carries 20 tons of fuel, flies in the stratosphere, cruises at Mach 2.5 in minimum afterburner and exceeds Mach 3.0 with ease when required – what can one say? It was an awesome aeroplane. The fact that the ventral fuel tank was one MiG-23 [equivalent in fuel] under the belly, speaks for itself.

Which words best describe the MiG-25?

Catch me if you can.

What is the cockpit like and how pilot-friendly is it?

Most Russian aircraft cockpits evoke a feeling of comfort and familiarity to a pilot who has flown Russian aircraft before. Coming from the MiG-21 to the MiG-25R was an easy transition. As one of our air chiefs remarked when the aircraft was demonstrated to him and he was stepping into the cockpit, 'This is rather familiar. And, dammit, it even smells the same!' The cockpit was a little more spacious than the MiG-21; thankfully so, because we operated wearing a pressure suit (which, incidentally, was the same as that worn by Yuri Gagarin).

The two-seater – or 'trainer' version – was unique; it is the only aircraft I know (other than the Tiger Moth, I guess) where the trainee sits in the rear seat. The design, to my mind, was an aeronautical engineering masterpiece. To put it rather simplistically, the camera block was removed from a single-seater and a cockpit created in that space. The canopy, although the same as the other cockpit, appeared 'flush' with the nose of the fuselage, as viewed from the rear cockpit. Thus, the trainee felt he was sitting in a single-seater when in the trainer. The transition to going 'solo' was a piece of cake. With the nose-

LEFT: As it was cloaked in secrecy, few photographs were taken of the day-to-day operations of the Indian MiG-25R. This is the two-seat MiG-25U (or 'trainer' version) – the rear seat is for the new pilot, the nose-mouthed cockpit for the instructor.

wheel located behind the rear cockpit, a 90-degree turn onto a taxi track entailed the front cockpit extending over onto the grass beyond the track (at the 'T') before the turn was executed. A little unnerving, initially, for anyone (though airline pilots may not have felt uncomfortable).

What can you say about the performance of the MiG-25?

It was a beast with immense power. It has been described by some as 'an engine with a place for a pilot and some avionics'. The Tumansky R-15B engines each provided more than 10 tons of thrust to produce the desired performance. In almost all the other aircraft I have flown, a regular climb is executed at constant TAS [True Air Speed: the speed of the aircraft relative to the air mass in which it is flying], with a progressive reduction of IAS [Indicated Air Speed: the speed as shown on the airspeed indicator] as the altitude increases. The Foxbat climbs at constant IAS with an increasing TAS, crossing abeam the take-off dumb-bell (if a reciprocal turn were to be executed after take-off) at 30,000 feet and increasing! She would be crossing 20 km [65,000 feet] in 6.5 minutes from

ABOVE: Operating at the edge of the Earth's atmosphere, MiG-25 crews required high-altitude pressure suits similar to those worn by cosmonauts. The high-altitude pressure helmet was the robust Model GSh-6A.

RIGHT: The Indian Air Force's Foxbats played a vital intelligence-gathering role. They were operated by 102 Sqn (the Trisonics) from 1981 to early 2003, and 35 Sqn (the Rapiers) from 2003 to the MiG-25's retirement in mid-2006.

wheels-roll, at a rate-of-climb of 100 m/sec [almost 20,000 ft/min] – like a bat out of hell, if you didn't come back from the max afterburner regime. In comparison, the ROC of a MiG-21 was 110 m/sec at sea level. Now, *that* is sheer performance. Cruising at 20-plus km with minimum afterburner (which, incidentally, provided best specific fuel consumption), she could execute a 45–50 degree bank-turn with just a wee bit of additional power. There was no loss of height. Her systems and autopilot were coupled to provide an optimised 'Little m=1' (remember the formula for maximum range?). So, as the fuel depleted, she would keep climbing (cruise-climb) and a mission commenced at, say, 19.5 km altitude would terminate around 22 km with no change of throttle position. The climb was so gradual over the period of time and distance that it did not affect the photography.

Were you detectable by radar? Were you susceptible to interception?

Certainly, we were detectable by radar, provided you were expecting us. The Foxbat operated covertly, seen just as a blip on the radar among other flying aircraft, but one blip would suddenly disappear. In normal ground radar settings the Foxbat generally operates at the highest fringes of the radar lobe, with

'The Foxbat operated covertly, seen just as a blip on the radar among other flying aircraft, but one blip would suddenly disappear'

the ingress and egress (through the radar lobe) often allowing one or two blips for the radar controller to perceive. Low transition times (because of the high speed) did not provide adequate reaction time to scramble fighters, and other than a pure head-on interception with look-up/shoot-up capability (from, say, 40,000 feet), the Foxbat could survive any fighter interception.

What were the limitations of the aircraft?

The fuel quantity, I guess. The engines were gas-guzzlers and 20 tons of fuel (including the ventral tank and fuel in the vertical fins) was just adequate. In the regional perspective of India and its neighbours it would suffice, but we always returned for landing with 200–400 kg of fuel remainder (200–250 kg was the

ABOVE: A two-seat MiG-25 shatters the peace. The Foxbat was a brutally effective solution to extreme high-speed flight utterly different to the US's sophisticated SR-71.

'In summer, with the ambient temperature close to 40°, the cockpit conditions with the canopy closed was a killer (start-up time to take-off being around 20 minutes)'

ABOVE: The thunderously loud MiG-25 shows off the ferociously afterburning Tumansky R-15B engines required to power the massive Foxbat. Each of these titanic engines generated over 10 tons of thrust.

fuel required to execute one circuit and landing). We operated on the fringe. The runway had to be kept clear at landing base (no other flying permitted for fear of runway blockage) once the MiG-25 commenced his descent. We needed to give only three RT calls – one for take-off, one for commencing descent and one for landing (in operational missions just two). There was no need to give any other RT calls because you operated unhindered in the stratosphere.

What does operating in the stratosphere feel like?

The subtle change in the colour of the sky starts around 16 km [50,000 feet], I guess, as the suspended particles which reflect/refract the sunlight start getting dissipated. The sky turns a distinct grey as you cross 20 km [65,000 feet] and continues getting darker as you transcend into those dizzying altitudes of 90,000 feet and 100,000 feet. You fly with cockpit lighting 'On' (as for night flying). It is a little eerie, one must admit. Not natural. The Earth is round, a fact we could confirm (!) because you can see the curvature of the Earth very clearly from those altitudes. The sun, moon, stars, and the illuminated ground below, are all visible at the same time. A glorious feeling.

All this was fine in winter, but in summer, with the ambient temperature close to 40°C [104°F], the cockpit conditions with the canopy closed was a killer (start-up time to take-off being around 20 minutes). Like other MiG aircraft, the heating system was brilliant, but the cooling system was designed to cut in only at one kilometre above ground level (and cut out at the same height during the return). Four layers of clothing – underwear, silk inners, pressure suit, flight suit – in those temperatures meant you were soaked to the skin by the time you returned to the aircrew room. It needed an extra effort by the trained assistant to peel the wet pressure suit and inners off your body. Guess we got our share of sauna baths!

What were your biggest fears when flying the MiG-25 – or were there none?

When you are flying a virtual fuel tank, the biggest fear is the illumination of the 'Fire' warning lamp. This is more so at operating altitudes in the stratosphere. Then there was the ejection seat in the MiG-25. This was the same as that in the later models of the MiG-21 and MiG-23. It had two settings – 3 km [10,000 feet] and 6 km [20,000 feet] – depending upon the terrain one was operating over. We set ours to 6 km. But operating at, say, 20–22 km altitude, where the

ambient temperatures are around minus 85°C, an ejection meant a free-fall of 15 km [50,000 feet] before the seat separated and activated the parachute. Would you hit terminal velocity? I guess you would. It was not a happy thought. The other fear was that, God forbid, one had to eject over enemy territory. On landing, how fast could one get out of the pressure suit (without external help) and be unbridled and unhampered to scramble for an escape? We practised and mastered the art in the squadron.

The high temperatures of the external surfaces of the aircraft, caused by travelling at great speed, also necessitated good cooling systems for the avionics and cameras.* This was achieved by alcohol (98 proof!); the MiG-25 consumes almost 200 litres of alcohol per mission. Alcohol bowsers [ground-based fuel trailers] were provided for replenishment, with a 'tap' provided at the rear – Aah! Don't you just love the Russians? – for purposes best left to your imagination! (Venting, perhaps.)

After the MiG-25, how did it feel to fly other fighters?

They were all sports models compared to the bulky MiG-25! Perhaps the greatest joy was to be able to throw the fighter around in the sky with gay abandon (which you missed when you flew the Foxbat), do aerobatics, fire weapons, and the adrenalin of doing air combat. We missed the G! Also, the sheer thrill of seeing another aircraft in the same sky!

Were you ever concerned about enemy defences? What actions would be initiated if you were painted on enemy radar?

It would be naïve of any warrior not to be concerned about his enemy. As I have said, missions were covert and silent. Just two RT

---

* Supersonic flight creates shock waves that rapidly compress air and lead to considerable heating.

calls – for take-off and landing. There were no warning systems in the aircraft. The only warning that could be given was by our own ground radar picking up a possible interception. Depending upon the threat, it would entail moving the throttle up the quadrant and initiating a gentle climb. Secrecy, speed and altitude were our only weapons.

Were there any aero-medical aspects that affected pilots flying this sort of aircraft?

We were subjected to aero-medical scrutiny for the first year of operation of the aircraft. There were two issues of concern to the doctors. Firstly, the extent of exposure to UV rays because of the clarity of the troposphere. We were made to carry dosimeters on our person; the results indicated there was no cause for concern. The second was the phenomena of possible psychological 'disassociation'. This came into consideration because of the rather lonely and silent missions in the troposphere, detached and distant from regular flight profiles. The issue was discounted because of the relative short duration of the missions – one hour at best.

Some parting words?

The MiG-25R was a superb flying machine, eminently suited to its task. It provided a feeling of immense power, invincibility and supreme confidence to the pilot in the execution of his mission.

BELOW: An MiG-25R of 102 Sqn 'Trisonics', Indian Air Force. Visible on its nose are the sensor ports required for the reconnaissance role. At high altitudes its 1,200-mm camera was said to be capable of photographing sites over 250 km away. This meant the safe observation of Pakistani areas of interest from Punjab or Kashmir.

THE UNSEEN MENACE

INTERVIEW WITH LOCKHEED F-117 NIGHTHAWK PILOT MAJOR ROBERT 'ROBSON' DONALDSON

The F-117 Nighthawk was a 'silver bullet', able to effortlessly penetrate the best-defended air spaces in the world by virtue of stealth, as first demonstrated to the world during Operation Desert Storm against Iraq. Major Robert 'Robson' Donaldson describes taking the stealth fighter to war.

THREE WORDS TO DESCRIBE THE F-117?

Deadly. Stealthy. Mysterious.

THE INVISIBLE MEN

The personality of the pilots chosen was actually a big deal since everyone was handpicked. In peacetime training, we were 200 miles north of Las Vegas at our base on the Tonopah Test Range, Nevada, for four to five days each week, so we all had to get along because we were in this same fishbowl. I think we were all of the same mindset, that even if the mission was a one-way suicide, we would still go. In wartime, I had to trust my fellow pilots with my life, and I did.

MEETING THE GHOST

I first saw the F-117 in a closed hangar at our base in Tonopah. Each jet had its own hangar that was shut closed during daylight hours to hide it from spying eyes and Soviet satellites. I walked into a dark hangar with my escort pilot and then the lights came on. There sat a futuristic spaceship with a large American flag hanging overhead. I was stunned. I thought, *That's not real. . . It must be a mock-up or something!* It defied all aviation logic.

I went over and touched the jet, and then started asking my escort pilot a lot of questions. We walked around the outside and I asked if I could sit in the cockpit. He said yes so I climbed up the ladder and sat in the seat. It was a very spacious cockpit compared to the F-16, which I had just come from. The stick was in the centre like a conventional fighter cockpit and I could see that the design was 'hands on throttle and stick' [HOTAS]. I asked about the various countermeasures the jet carried and his answer to all those questions was: 'Not needed.'

I then closed the canopy and it felt like I was closing my coffin lid. The visibility out the front was very poor because of the location of the HUD [head-up display], and there was no rearward visibility; the side windows were really the only place to see out (very unlike the almost-unlimited visibility I enjoyed in the F-16). I was not yet in class to learn how to fly and tactically employ the jet, but I was really excited to be a part of the programme and get started learning.

TAKING THE 'BATPLANE' TO COMBAT

We had a tremendous amount of confidence in the capability of the jet to slip through the Iraqi integrated air-defence system and precisely deliver our munitions on target on time. The F-117 had been extensively tested against all aspects of a Soviet IADS [integrated air defence system] in the USA. Also, several times during the Desert Shield build-up we ran multiple F-117s right up to the border with Iraq to see what the Iraqi reaction would be: no response was detected by the assets that collected ELINT

OPPOSITE: The strange alien form of the F-117 is less about efficient aerodynamics and more about the control of incoming radar energy. The angles of the aircraft's faceted surface panels are carefully considered to ensure little of the radar signals is reflected back directly to the enemy transmitter/reciever.

LEFT: The extreme wing sweep keeps radar frontal cross section to a minimum and deflects incoming radar waves. As a pioneer of stealth, the F-117 sacrificed both handling characteristics and performance in pursuit of the minimum possible radar signature. Today all new fighters and attack aircraft are designed with stealth in mind.

'Underneath that black cloud was an absolute Dante's *Inferno* scenario – a sight I'll never forget'

[electronic intelligence] so we positively knew we could slip by them undetected. Our only concern as pilots was the 'golden BB', a random bullet of AAA [anti-aircraft artillery] that would hit our jet and take us down. With vast amounts of bullets from 23-mm, 37-mm, 57-mm, 85-mm and 100-mm guns, this was a very real danger. Iraq was a fully armed nation and the tremendous amount of AAA shot up was the same on the last night of war as it was on the first night.

My most memorable mission during Desert Storm was when I was tasked to destroy two bridges in south-east Iraq, close to the border with Kuwait, in anticipation of the ground-assault phase. The significance of the bridges was obviously the Iraqis' ability to resupply their troops in Kuwait. After a top-off of fuel from the KC-135 tanker near the Saudi–Iraqi border, I entered Iraq in our 'stealthed-up mode' (all external lights off, no comm., no antennae extended). Saddam Hussein had already set the oil wells on fire in Kuwait, so as I approached the area I could see there was a heavy

black cloud from those oily fires obscuring the ground. My profile called for bomb release somewhere around 12,000 feet but since our rules of engagement required a positive identification of the target, I knew that would not be possible from that altitude. I did not want to return to my base with those two GBU-10 bombs (2,000 lb each with a laser-guided kit). Knowing the terrain is flat in that area, I decided to descend to get below the oily cloud-layer so as to be able to positively ID the bridges, which were about seven miles apart from each other. I had to get down to about 700 feet above ground level in order to be in the clear.

Underneath that black cloud was an absolute Dante's *Inferno* scenario – a sight I'll never forget. I went over my IP [initial point: an easily identifiable feature used as a starting point for the bomb run] and lined up on the first bridge. Visibility was actually quite good because the oily overcast reflected the fires light underneath. I pushed up the throttles to go as fast as I could and my inertial navigation system had positioned the cross hairs of my laser sight right on the bridge, so visual ID of the target in my cockpit was accomplished. I aimed for the far end of the span on the bridge. The single bomb released very close to the target and I ripped the throttles to idle to slow my speed so that the laser would not gimble – low to the ground and fast, the laser would run past the span before my bomb hit, thus rendering it a dumb bomb). Ripping the throttles to idle caused the jet to decelerate so that I

MISSIONS AND WEAPONS OF WORLD WAR THREE

If a Cold War scenario erupted, we would have been flying out of a base in the UK. The first night missions would probably be a combination of striking high-value targets such as communication nodes, integrated air defence assets and other strategic targets, along with hunting and shooting down a Soviet AWACS [Airborne Warning and Control System]. If the decision was made to use nuclear weapons, we were fully prepared to deliver those also.

The Anti-AWACS role was really just a concept, as all aspects of the capabilities of the stealthy jet were examined. It was talked about very briefly in our classroom instruction but we never flew a practice sortie. The concept was tested early on by the small cadre of test pilots using AIM-9L infrared heat-seeking missiles, as they were the only air-to-air missiles we could carry. Presumably we would have carried two missiles to use against one Soviet AWACS aircraft. I think the air-to-ground missions would have been more dangerous than hunting the AWACS because more threats existed in that arena (AAA, SAMs) than in the air-to-air arena.

We could carry every weapon in the US Air Force inventory (except a radar-guided AIM-7 Sparrow air-to-air missile). When I flew the jet, we had an inertial navigation system from a B-52 that took 43 minutes to align. It was super-accurate for its time but obviously the F-117 was not a quick-reaction/scramble jet. GPS was not yet available but was retrofitted to jets several years later. Lockheed flight-tested a radar in place of the FLIR [forward-looking Infrared] but decided against putting it in the jet. GPS-guided bombs would have been nice to have then because we would not have had to weather-abort a mission because we could not visually identify the target; we could have just released a GPS-guided bomb above the weather with a high degree of confidence that it would hit the target. But all in all, for its time, the F-117 had the best sensors and weapons of any aircraft out there.

The greatest myth about the F-117 was that it was a completely invisible: in reality, it was a Very Low Observable [VLO] jet. Certain aspects of how we flew against enemy radar (altitude, route, aspect) was factored into our mission-planning in order to maximise our VLO capability.

had time to keep the laser beam on the span but it also caused my head to tumble so that I felt like my head was going end-over-end through space. The bomb impacted the bridge and the explosion caused the span to drop into the river. A nanosecond later I saw on my screen an Iraqi army truck drive off the bridge where the span had been a moment before. I had no time to process that snapshot as my jet was rocked by the explosion of my own bomb, turning me about 135 degrees upside down and disengaging the autopilot. I managed to recover the jet about 400 feet above the ground, climb back up to 700 feet and re-engage the autopilot. The fragmentation envelope of a 2,000 lb bomb is 2,500 feet in all directions upward and to the sides of the impact, so from 700 feet above the ground I was well within the frag envelope of my own bomb. I knew that but had decided to take my chances anyway.

Once I was right-side up, I immediately looked at my engines to confirm they were running and to check my fuel status. Both were good so I didn't think I had fragged myself. So a lot was happening in a very short period of time and now I had to line up on my second bridge, which was rapidly approaching, but this time I knew what to expect. Once again, the deceleration, the tumbling, the explosion, a dropped span, and being blown upside down were all that I had anticipated, only this time I climbed up after I recovered the jet and headed for the air-refuelling point. I carefully looked over the engine instruments and my fuel status, and once again all seemed normal. After I had post-strike air-refuelled, I had about 90 minutes to reflect on that sortie. It was very satisfying to drop those two bridge spans but I wasn't 100 per cent sure that the Iraqi Army truck driving off the span actually happened or if it was my mind playing tricks on me. I'm also pretty sure that since the

bombs I dropped had a slight delay-fuse, the bombs penetrated the concrete and then detonated just under the bridge so that a good portion of the bomb fragments would have been trapped underneath the bridge and not gone up in the air into my jet. I landed back at base and our routine was to look the jet over on the outside after each mission and make sure all was intact. So I told my crew chief I was a bit concerned but we both looked the jet over carefully and did not find any self-inflicted holes. We have a saying in the fighter community that God takes care of dumb farm animals and fighter pilots. I'm not sure which group I'm in!

Next was a stop and debrief with our intelligence people to assess the strike via the VHS camera film. Sure enough, right there on film was the Iraqi Army truck driving off the span. All in all, it was a very surreal night and I was happy to finally crawl into bed and let my brain and body get some sleep after a five-and-a-half-hour mission.

LEFT: The F-117's usual reticence was subverted for this patriotic scheme around the time of the type's official retirement. It was later revealed that the Nighthawk had never retired and was again living in the shadows, this time as a stealthy aggressor aircraft for testing and training purposes.

THE FREUDIAN
GUIDE TO SPY PLANES

Schaulust is a German word that describes the pleasure we feel in looking. This curiosity, or scopophilia, is a powerful pleasure and one that, if not indulged in directly, will lead to sublimation in another activity, often one remarkably similar to erotic voyeurism. The love of a military aircraft finds its roots in the purified pleasure-ego, that facet of the ego that projects its own badness onto external objects.

'The spy plane is an example of the hyper-sexualisation of the object (often bolstered by patriotism, itself a product of the *Über-Ich*)'

The spy plane, exemplified by the erotically supercharged SR-71, is an example of the hyper-sexualisation of the object (often bolstered by patriotism, itself a product of the *Über-Ich*). Sitting at such a giddy vertex of powerful impulses makes the spy plane a powerful symbol. It is because of this that the spy plane occupies such a particularly potent appeal to the unconscious.

Let us start with the Boeing E-6 Mercury. It is surely significant that this aircraft was originally known as the Hermes – Hermes, the god of 'boundaries', is a phallic deity. In ancient Greece he was portrayed on pillars with an engorged phallus – his son Pan was shown with a larger and fully erect penis.

The E-6 is codenamed 'Looking Glass', as this is not a spy plane but a mirror to the US Navy's ability to control forces when a nuclear attack destroys Global Operations Center (GOC), located at Offutt Air Force Base. The mirror also reveals a nation in the grip of the most dangerous of anti-cathexis – the belief in nuclear deterrent; the unconscious desire for a universal orgasm that will consume and destroy all of us.

SR-71 BLACKBIRD PILOT B. C. THOMAS EXPLAINS WHY WE SHOULDN'T SAY 'SPY PLANE'

The SR-71 was a 'strategic reconnaissance aircraft', not a 'spy plane'. The practice of calling the SR-71 a 'spy plane' is so prevalent that I have stopped trying to correct the error, and it is no longer important since the SR-71 is no longer flying, although the U-2 pilots have cause to resent their being called 'spy pilots'. We have resented that moniker because of the formal, international consequences of being captured as a spy, as opposed to a military man flying a marked military aircraft while wearing a military uniform with name and rank displayed, and carrying a military identification card which is also a Geneva Convention identification card. Our status as a military pilot on a military mission was supposed to carry with it certain prerogatives which other countries were 'constrained' to recognise, but whether they did or not was another question. Routinely, spies are summarily executed; military men captured are supposed to be treated in accordance with the Geneva Convention. Big difference!

'Our status as a military pilot on a military mission was supposed to carry with it certain prerogatives which other countries were "constrained" to recognise'

FEHLLEISTUNG

The enemy's aircraft has the same symbolic power as our own, but its value is inverted. That the Mikoyan-Gurevich MiG-25R was made so dangerous to fly was a nod to Freud's 'death drive', the fundamental tendency for life to seek the calm of entropic non-activity. A *Fehlleistung* is a faulty achievement – when we intend to do one thing but do another – and is really the code word that should have been assigned by the Air Standardisation Coordination Committee (ASCC) to describe this extremely fast aeroplane. The MiG-25R was virtually invulnerable to interception (at least physically, though not emotionally), relying on high speed and altitude to evade its persecutors.

Throughout the 1970s, it suffered a repetition compulsion to fly over troubled Middle Eastern nations without offering help. Its attraction to potential conflict continued through its later life – in 1997 a MiG-25R of the Indian Air Force caused a diplomatic stir by snooping on Pakistan at bi-sonic speeds. If the Foxbat embraced *Thanatos* (self-destructive death instincts) then the Blackbird represents the opposite: it is the voice of the libido and of the *Trieb* (urge and desire) – the invincible, sexual self.

RETRACTIONS

The retractable air-refuelling probe used by some non-USAF aircraft was developed as a piece of kinetic sculpture exploring *koro*, a widespread delusion that one's penis is retracting and will disappear. The now very tired witticism comparing air-to-air refuelling to sex was declared a UNESCO World Heritage Joke in 2009.

But what of the female? The Lockheed Electra complex is the desire of the female child to have children with her father in a four-engined turboprop. So taboo was such a notion that the Apollonian US Navy disguised the Electra, renaming it Orion (a name with altogether different associations) when it pressed it into service, notably as the EP-3 version, which – revealingly – was a reconnaissance aircraft.

THE BIRTH OF THE UNMAN

The unmanned aircraft – its title alone expressing the agonising castration-fear of pilots – is a nightmare woven from carbon fibre. Who controls what it sees? The Latent Control parallels are as dangerous as you would imagine. With its famed 'drinking-straw vision', does it demonstrate a wilful aversion to distinguishing between inner and outer life? Can it separate fantasy and externality for its Portakabin dreamers? The unmanned aircraft is voyeurism heaped upon voyeurism – a remove from a titillating remove, and as such threatens all involved.

Sigmund Freud's new book *Harrier GR.3: The Thimble-Nosed Mother of All of Us* is now available.

ABOVE: Freud believed that the mythical Medusa's head represented female genitals, with the writhing snakes symbolising pubic hair. The prototype of the MiG-25, designated Ye-155, first flew in March 1964. Unlike the later operational aircraft, it included tip-tanks and winglets. The absence of these features has a clear analogous link to Freud's Medusa.

DECLASSIFIED

INTERVIEW WITH SR-71 BLACKBIRD PILOT B. C. THOMAS

From the 1960s until the 1990s the US spied on whomever it liked with impunity, using the snapping cameras and greedy sensors of the fastest aeroplane ever to take off from a runway: the spectacular SR-71 Blackbird. Pilot B. C. Thomas spoke to Hush-Kit about life in the most exciting seat in the world.

What was the best thing about flying the SR-71?

The best aspect was the mission, and I believe all who supported or flew the aeroplane operationally would agree. I was absolutely thrilled to be part of the strategic reconnaissance effort of the United States and, by extension, the Free World, to survey our potential enemies and glean information that only we could provide, owing to our reconnaissance capability (sensors), and our stealth, flexibility, speed and altitude. We advertised, that with 24-hour notification, we could be over any spot on Earth and capable of revealing what was there. That boast was successfully tested many times. And to a pilot who actively sought excitement paired with meaningful accomplishment, the notion of flying the fastest and highest-flying aircraft in the world while contributing to national security was unbeatable.

. . .and the worst thing?

The worst part of flying the SR-71 was the environment in which we flew. We flew fast and high, which complicated controllability and made over-controlling very dangerous because the SR-71 was delicate and not very manoeuvrable, compared to other high-performance fighter aircraft. At Mach 3 and above, which were our usual cruising speeds, our acceleration limit was only 1.5 G, or 45 degrees of bank, because of structural heating. We also operated in near-vacuum, where the air pressure was about 0.4 pounds per square inch, and if we were unprotected, our blood would boil and death would be instantaneous. To achieve enough dynamic air pressure to sustain lift, we had to fly fast, when air friction caused the average skin-temperature of the aircraft to be 600°F. The afterburner section was over 1,200°F. We cruised at 15 miles above the earth, so any cockpit environmental problem, such as high temperature, low pressure or oxygen depletion, could be fatal, because slowing down and descending could not be achieved quickly.

What is the highest and fastest you've flown in the SR-71?

At Beale Air Force Base, I flew no faster than Mach 3.25, and while testing new systems and equipment in the SR-71 at Edwards Air Force Base, we flew almost all missions at Mach 3.2, which was the highest Mach that was attained on the vast majority of operational missions. But for some test flights, like testing the digital automatic flight and inlet control system (DAFICS), we tested the full flight envelope to Mach 3.3, which is the fastest I flew the Blackbird.

The highest altitude I reached was 86,000 feet, while flying a Murmansk mission [during the Cold War Murmansk was a centre of Soviet submarine activity]. I had to fly that high so that I could keep the speed at or below Mach 3.2 – my target speed – while in minimum afterburner. We were never power-limited and most high-Mach cruise missions were flown with the throttles below half-travel within the afterburner range.

BELOW: 'That is me in front and Phil Soucy in the back seat. It was taken just before I took off at Norton AFB, CA, to fly in an air show in 1986.'

negatively impressed them with our surveillance flights, that they knew we were there, and there was very little they could do about it except write such obvious and typical Communist propaganda screed. We had a few laughs and a round of cheer was in order.

Very little was heard from them until 25 August 25 1981. Pilot Maury Rosenberg and RSO Ed McKim were flying a two-loop (our moniker for a mission involving two refuellings) reconnaissance mission, first against Communist China and then North Korea. The pass across North Korea was along the demilitarised zone between North and South Korea, although North Korea claims sovereignty over South Korea as well. On the second pass, the North Koreans launched a surface-to-air SA-2 missile in an attempt to shoot down the SR-71. They missed by several miles.

Jay Reid and I were at RAF Mildenhall when this happened, and we pilots and RSOs were given a detailed briefing about the incident. Our reaction was not to be very concerned about their ability to hit us, but speculate what change it might portend for future missions. Perhaps we would fly deep into North Korea's territory, fly more often at night or increase our sortie rate. In any case, we figured that something would change as a result of their belligerence.

About a month later, on 24 September 1981, Jay and I arrived in Okinawa to start a regular six-week deployment. Two days later, the deputy secretary of defense, Frank Carlucci, came to our detachment (Det 1, 9 SRW) to inspect our SR-71 operations. Since Jay and I had been on the island only two days (we were not allowed to fly until we had been jet-lagged-acclimated for three days), we were designated his briefing officers specifically to show him our airplanes and answer all questions he might have. Part of our briefing included showing him the SR-71, putting him in both cockpits and giving him an overview of our mission procedures. We especially emphasised the unusual aspects of the aircraft, including the unique controls for the engine inlets, and the defensive and navigational systems. He expressly asked us about the pilots' and RSOs' attitude about flying operational missions, especially in light of the attempted shoot-down. We assured him that we were all dedicated to those missions and that the prospect of another missile attack did not particularly bother us because we had ultimate faith in our defensive equipment and our ability to manoeuvre. Carlucci specifically stated that President Reagan was 'furious' that the North Koreans had fired on one of our aircraft and that something would be done about it. In the meantime, we were to fly our reconnaissance missions 30 miles south of our normal flight paths.

'The prospect of another missile attack did not particularly bother us because we had ultimate faith in our defensive equipment and our ability to manoeuvre'

What was your most memorable mission – and why?

That was when the consequences of one particular flight may have started a war. The background for this flight began on 13 November 1980, when RSO [reconnaissance systems officer] Jay Reid and I flew a mission against North Korea. This was just after President Reagan was elected and North Korea was sending a message to the new, incoming administration that our flying reconnaissance near or over their territory was unacceptable. The Communists sent this message the next day, specifically mentioning our flight. Obviously, the North Koreans were not happy about our flights.

We SR-71 crew members thought it was great to receive such a tirade from the North Koreans. We knew that we had

ABOVE: B. C. Thomas (*left*) with RSO Jay Reid.

Ten days later, on 3 October 3 1981, the US Air Force vice chief of staff, General Robert Mathis, came to Okinawa and briefed the SR-71 crews on the plan to resume normal operational flights. He said that soon we would fly a mission exactly like the one flown when the missile was launched at Maury and Ed. He said also that the timing would be critical and that we had to be over the North Korean missile-launch point within one minute [of the planned time], although we should be within 30 seconds if possible. He emphasised the timing was important because if the North Koreans fired another missile at us, US Air Force fighter aircraft would launch an air-to-ground missile attack on the North Korean launch site immediately.

Jay Reid and I flew that mission on 26 October 1981. We took off early, used 'timing triangles' to refine our time-over-target and passed over the launch site within 10 seconds of the critical time. We took a great deal of pride in successfully flying that mission as planned and in making a very strong statement that we, and by extension the United States, would not be deterred.

The North Koreans did not fire at us, and I'll admit that I was a little disappointed, for our reaction would have certainly demonstrated our national resolve. And I don't like Communist governments! Obviously smitten by our flight, and perhaps trying to bluster their way out of an embarrassing situation, the North Korean Communist government issued yet another propaganda blast. This is the message:

SK141394 REF SK140322 FYI

PYONGYANG KCNA IN ENGLISH 0353 GMT 14 NOV 80

   ((TEXT)) PYONGYANG NOVEMBER 14 (KCNA) -- THE U.S. IMPERIALIST
AGGRESSION FORCES WHICH ARE NOW HASTENING PREPARATIONS
FOR A WAR, FRANTICALLY KICKING UP A WAR DUST AGAINST
THE NORTHERN HALF OF THE REPUBLIC, COMMITTED A MILITARY
PROVOCATION BY INFILTRATING AN "SR-71" HIGH-SPEED, HIGH-
ALTITUDE RECONNAISSANCE PLANE AT AROUND 12:23 ON
NOVEMBER 13 DEEP INTO THE AIR ABOVE THE COASTAL SEA
OF OUR COUNTRY EAST OF KOSONG ON THE EAST COAST FOR
ESPIONAGE.
   WHILE RAISING A HUE AND CRY OVER FICTITIOUS "THREAT
FROM THE NORTH" THESE DAYS, THE U.S. IMPERIALISTS
PERPETRATE MILITARY PROVOCATIONS MORE FREQUENTLY AGAINST
OUR PEOPLE, ZEALOUSLY ENCOURAGING THE CHON TU-HWAN
MILITARY FASCIST GANGSTERS TO AN "ANTI-COMMUNIST"
SMEAR CAMPAIGN.
   AFTER SEPTEMBER 4 ALONE, THE UNITED STATES HAS
INFILTRATED SPY PLANES ON 36 OCCASIONS INTO THE AIR
ABOVE THE TERRITORIAL WATERS AND COASTAL SEA OF OUR
COUNTRY.
   IT IS A CHALLENGE TO PEACE IN KOREA AND HER PEACEFUL
REUNIFICATION AND A DANGEROUS ACT JEOPARDIZING PEACE IN
ASIA AND THE REST OF THE WORLD THAT THE U.S. IMPERIALISTS
CONTINUE PERPETRATING ESPIONAGE ACTS AGAINST OUR PEOPLE
AT A TIME WHEN OUR NEW MOMENTOUS PROPOSAL TO GUARANTEE
PEACE IN KOREA AND ACHIEVE HER PEACEFUL REUNIFICATION
HAS EVOKED UNANIMOUS SUPPORT AND SYMPATHY OF THE ENTIRE
KOREAN PEOPLE IN THE NORTH AND THE SOUTH AND THE WORLDS
PEACELOVING PEOPLE.
   SUCH FREQUENT ESPIONAGE ACTS OF THE U.S. IMPERIALISTS
TODAY CLEARLY SHOW HOW RECKLESSLY THEY ARE RUNNING TO
START A NEW WAR OF AGGRESSION IN KOREA.
   THE U.S. IMPERIALISTS MUST IMMEDIATELY STOP THEIR

We didn't follow the Communists' advice and our reconnaissance missions against North Korea continued unabated.

Another significant mission for me and Jay Reid was the time an aircraft emergency forced us to land unannounced in Continental Europe (specifically Norway) with highly classified mission materials in the SR-71.

LEFT: An SR-71 lands with its drogue chute deployed. Every single SR-71 flight was a meticulously planned and executed operation more akin to space travel in terms of effort than most other forms of flying.

TOP 10
WORST BRITISH AIRCRAFT

If you want something done slowly, expensively and possibly very well, you go to the British. But while Britain created the immortal Spitfire, Lancaster and Edgley Optica, it also created a wealth of dangerous, disgraceful and diabolical designs. These are just ten plucked from a shortlist of thirty. In defining 'worst', we've looked for one, or a combination, of the following: design flaws, conceptual mistakes, being extremely dangerous, being unpleasant to fly, or being obsolescent at the point of service entry (and the type must have entered service). Grab a cup of tea and prepare for ire as you read about ten machines they wanted your dad, grandad or great-grandad to fly to war.

10. Blackburn Beverley (1950)
'The Beverley hellbilly'

A mere year separates the service entry of the Beverley (1955) and the US's Lockheed C-130 Hercules (1956), yet sixty years later one of these aeroplanes is still the best tactical transport, serving with many air forces around the world, and the other exists only in the form of a single lonely airframe converted into a B&B. There's a reason for this.

The Beverley had four Bristol Centaurus engines, together capable of generating 11,400 hp, pulling a fully-loaded Beverley weighing 135,000 lb (61,200 kg); the C-130A had a maximum weight of 124,200 lb (56,250 kg) and had 15,000 turboprop horsepower to move it. The Centaurus also powered the abysmal Blackburn Firebrand, the pitiful Bristol Buckingham and the technically brilliant (but conceptually wrong-headed) Bristol Brabazon – and, for the sake of fairness, the superb Hawker Sea Fury. Lockheed threw vast resources at getting the Hercules right (so much so that the great engineer Kelly Johnson thought the project would sink the whole company), whereas Blackburn used warmed-up Second World War technology and a dawdling development time to produce an aircraft that was at best mediocre and that did its own small part in teaching the world that America was better at making aeroplanes.

In defence of the Beverley, it performed well in austere conditions and could be procured without spending foreign-currency reserves.

OPPOSITE: The Supermarine Swift was a deeply flawed fighter but was saved from this list by its more successful career in reconnaissance.

ABOVE: Is it possible to waddle in flight? If so, then the adorable but questionable Beverley would indeed have waddled.

9. Supermarine Scimitar (1956)
'Red Beard's scabbard'

Take an aircraft so dangerous that is statistically more likely than not to crash over a twelve-year period – and arm it with a nuclear bomb. Prior to this, ensure one example crashes and kills its first commanding officer, in front of the press. There you have the Scimitar. Extremely maintenance-heavy, an inferior fighter to the Sea Vixen and a worse bomber than the Blackburn Buccaneer, the Scimitar was certainly not Joe Smith's finest moment: his more glorious achievement was leading the development of the Spitfire from 1937 onwards. It was the last Fleet Air Arm aircraft designed with an obsolete requirement to be able to make an unaccelerated (without catapult) carrier take-off, and as a result had to have a thicker and larger wing than would otherwise be required. Only once did a Scimitar ever make an unassisted take-off, with a very light fuel load and no stores, and then just to prove that it could be done.

'Take an aircraft so dangerous that is statistically more likely than not to crash over a twelve-year period – and arm it with a nuclear bomb'

8. Panavia Tornado F2 (1979)
'The timcat'

The Tornado interceptor was a very British development of an international aircraft. In the 1970s, the British Aircraft Corporation (BAC) pushed heavily for an interceptor variant of the Tornado (a ground-attack aircraft). The Government and partner nations were sceptical that this project would be the low-cost, low-risk, high-performance fighter promised, so BAC massaged the facts a little, deliberately understating what a huge undertaking it would be. Essentially they took a heavy airframe optimised for low-level flight, with engines also optimised for low-level flight and a radar optimised for attacking ground targets from low-level flight, and attempted to turn it into an interceptor intended to attack bombers at medium and high altitudes. To add to the fun, it was decided to develop an extremely ambitious new radar, despite Britain not having created an advanced fighter radar since the Lightning's fifties-technology AI.23 (the Sea Harrier's Blue Fox radar was a low-performance set derived from a helicopter system). Despite its 'F for Fighter' designation, and the euphemistic 'interim' description, the F2 ended up without a functioning radar, as the new sensor suffered terrible teething problems. Issues with the aircraft's centre of gravity, caused by the absent radar, were solved with a large chunk of concrete ballast satirically dubbed the 'Blue Circle radar' after a cement brand (the nature of this ballast was apocryphal – it actually comprised steel bars). Despite the Tornado's terrible high-altitude performance and poor agility, huge amounts of money and time led to the F3, which eventually matured into a capable weapon system. Quite how many F-15Cs could have been bought for the cost of the Tornado Air Defence Variant programme is a question many RAF crews moaned to themselves as they struggled to refuel at altitudes higher than the Post Office Tower.

ABOVE: At a certain angle and in certain light, the Tornado ADV was extremely good-looking, which would have been handy for charming its way out of a dogfight with a Su-27 Flanker.

7. Gloster Javelin (1951)
'It's not time for T'

It takes a special kind of genius to make an aircraft subsonic that has a delta wing and one of the highest thrust-to-weight ratios of its generation, but that's what Gloster did. The Javelin entered service in 1956, the same year as the dreadful Convair F-102, but even the disappointing American fighter would have smashed the Javelin in a drag race. Bizarrely, the use of afterburner below 12,000 feet (3,657 m) actually slowed the aircraft down! It visibly burned fuel but reduced the core rpm, therefore thrust; the engine core was competing with the reheat for the fuel-flow available. Add to this lamentable low-level instantaneous turn-rates, nightmarishly bad access to internal parts for maintenance crews and frequent post-start fires. After a mere twelve years in service, the RAF dropped the type. Unsurprisingly, no export orders were received for the 'Tripe Triangle'.

BELOW: According to Javelin pilot Peter Day, the aircraft was an excellent drinks cooler: 'A complete box of Tiger beer would fit into each gun magazine and be perfectly cooled after flight.'

6. Blackburn Firebrand (1942)
'Fleet Evil'

The story of the Firebrand torpedo fighter is a rotten one. The specification for the type was issued in 1939, but it wasn't until the closing weeks of the Second World War that it began to enter service. Despite a luxuriously long development, it was an utter pig in the air, with stability issues in all axes and a tendency to lethal stalls. There was a litany of restrictions to try to reduce the risks, including the banning of external tanks, but it still remained ineffective and dangerous to fly. Worse still, instead of trying to rectify the problems, the Fleet Air Arm (FAA) started a witch-hunt of those pilots who dared to speak the truth about the abysmal Firebrand. Only two Firebrand squadrons formed, of which the flying complement was heavily, if not entirely, made up of qualified flying instructors, suggesting only the most experienced pilots could be trusted with this unforgiving monster.

> 'Only two Firebrand squadrons formed, of which the flying complement was heavily, if not entirely, made up of qualified flying instructors'

5. de Havilland Sea Vixen (1951)
'Vixen Vapour Rub'

The Royal Navy's Sea Vixen fighters were deathtraps. One hundred and forty-five Sea Vixens were built, of which 37.93 per cent were lost over the type's twelve-year operational life, and more than half of those incidents were fatal.

The Sea Vixen entered service in 1959, a full eight years after its first flight and two years later than the US Navy's Vought F-8 Crusader, which was more than twice as fast as the Sea Vixen, despite having 3,000 lb less thrust. The development of the Sea Vixen had been glacial. The specification was issued in 1947, initially for an aircraft to serve both the FAA and the RAF. The DH.110 prototype first flew in 1951, but one crashed at the Farnborough Airshow the following year. This slowed down the project, which was then put on hold as de Havilland and the Royal Navy focused on the alternative, the DH.116 'Super Venom'. Once the DH.110 became prioritised again, it was substantially redesigned to fully navalise it. Then, when the Royal Navy gave a firm commitment, it requested a radar with a bigger scanner and several other time-consuming modifications. All of which meant it arrived way too late to be a world-class combat aircraft. Its peer, the F-8, remained in frontline service until 2000; its other contemporary, the F-4, remains in service today. The Sea Vixen retired in 1972, by which time fifty-one Royal Navy aircrew had been killed flying it.

ABOVE: The unlucky observer sat below and to the right of the pilot in what London estate agents would refer to as a spacious luxury living area: it was a cramped, dark 'coal hole' notoriously difficult to escape from.

'The prototype Lerwick took shape during 1937 and testing began in November that year as the first batch of ten were being constructed. Results were not encouraging'

4. Saro Lerwick (1938)
'Fat Boy Swim'

The Saro Lerwick closely resembled a scaled-down, twin-engine Short Sunderland but the dismal nature of its service career was in inverse proportion to the success of its larger colleague. Perhaps the most unfortunate of British aircraft, the Lerwick's persistence in active service, for three years, serves to show just how desperate Coastal Command was for any kind of aircraft, even dangerously ineffectual ones, during the early war years.

Alliott Verdon-Roe was the first person to build and fly a British aircraft with a British engine, achieving this milestone in 1909 with his Roe I Triplane. With his brother Humphrey, Verdon-Roe also founded the company Avro, which would become famous thanks initially to the hugely mass-produced 504 and later to the stalwart Lancaster. Verdon-Roe decided to leave Avro in 1928 and bought the S. E. Saunders boatbuilding company, renaming it Saunders-Roe, or 'Saro' for short, intending to develop flying boats. And Saro certainly developed a lot of flying boats – fifteen different designs were flown between 1920 and 1952, but they never managed to build more than thirty-one examples of any of them, the most successful being the somewhat humdrum biplane Saro London, which the Lerwick was intended to replace. A. V. Roe was prolific in producing aircraft of varying success for the British armed forces, so it may come as a surprise to learn that Alliott Verdon-Roe himself was a Fascist.

Verdon-Roe joined Oswald Mosley's British Union of Fascists in 1936 and this may explain the horrible Lerwick and its effect on both the RAF and the Royal Canadian Air Force. You could be forgiven for thinking that designing an aircraft to fly around slowly for long periods of time in the hope that someone might see a submarine and then drop something on it might be a relatively simple task, but the Saro Lerwick serves to prove that, apparently, it is not. Twenty-one were built and fourteen were lost (thirteen in accidents; one simply disappeared).

The whole unhappy Lerwick saga can be traced back to the Air Ministry's specification R.1/36, issued in March 1936, calling for an all-metal flying monoplane boat to replace the Saro London and Supermarine Stranraer biplanes, followed by the Ministry's unusually dynamic decision to order the aircraft 'off the drawing board' rather than wait for it to actually exist and be tested. To be fair, the Lerwick at least appeared to be a safer bet than the other two contenders for R.1/36. Supermarine had put forward a typically elegant aircraft but it was deemed too expensive and, anyway, the company were up to their neck in work, developing

the Spitfire. Meanwhile, Blackburn had come up with the fairly crazy B-20, which featured a semi-retractable hull and two untested and unconventional Rolls-Royce Vulture engines. By contrast, the Bristol Hercules-powered Lerwick looked decidedly sensible. Further confidence was engendered by the design team: chief designer at Saro was Henry Knowler, an engineer of great experience stretching back to 1915, and the Lerwick was the first aircraft that the promising young designer Maurice Brennan would work on. Brennan would later distinguish himself by designing a rather more impressive maritime patrol aircraft, the Hawker Siddeley Nimrod.

The prototype Lerwick took shape during 1937 and testing began in November that year as the first batch of ten were being constructed. Results were not encouraging. The aircraft's main problems were simple lack of power coupled with an inexplicable lack of stability both in the air and on the water. It couldn't be flown hands-off, which is problematic for a long-range patrol aircraft, and it couldn't maintain height on one engine. Even if it could, it would have been impossible to counteract the torque of the remaining engine, rendering it effectively impossible to continue flying in the event of engine failure. It was also prone to porpoising on

landing and take-off, and possessed a vicious stall. Add structural failures and a woefully unreliable hydraulic system, and it became painfully obvious that the Lerwick was not entirely suitable for its intended role. Modifications were made to the hull and tail in an attempt to ameliorate the worst aspects of the design, but to no great effect. The Lerwick was pressed into service due only to the lack of any immediate alternative and despite the December 1939 recommendation, made by Air Vice Marshal Sholto Douglas, then assistant chief of the Air Staff, to scrap the aircraft and have Saro build Sunderlands instead. Inexplicably, production continued, albeit at a snail's pace, until March 1941, when the twenty-first and last Lerwick rolled out of Saro's factory on the Isle of Wight.

As an omen of what was to come, on 1 September 1939, the first day of the Second World War, LZ249, the second Lerwick to be built, sank at its mooring at Felixstowe due to a hatch being left open at night. The RAF's 240 Squadron were the first to receive the Lerwick but in October 1939 they passed their aircraft on to 209 Squadron, who would become the sole squadron to operate the aircraft in its intended role, to patrol over the North Sea. Despite not yet serving on operations, 209 had already lost one Lerwick – on 20 February 1939, LZ7253 suffered a hard landing at night,

ABOVE: The bloody awful Lerwick has been described as the 'unlucky man's Sunderland'. At sea and at war, it was an utterly untrustworthy friend to its hapless crews.

causing the right-hand float to break off and the whole aircraft to sink, drowning four of the crew. Subsequently raised, LZ7253 demonstrated its desire never to fly again by sinking for a second time, this time permanently, during a storm.

About a month after this unfortunate accident, the Lerwick saw its first combat action. On 25 March 1940, Lerwick L7256 WQ-V, piloted by Flight Lieutenant Bennett, attacked an enemy submarine by dropping bombs, with no visible result. And thus ended the combat career of the Saro Lerwick flying boat.

The losses continued nonetheless. On 29 June 1940, LZ261 was lost when the right-hand float again broke off in a hard landing. That winter, L7251 and L7255 sank at their moorings on Loch Ryan. In January 1941, LZ262 lost control on take-off and crashed; in February LZ263 disappeared; in March LZ252 sank after one of its engines tore off its mount; and in October LZ254 sank after striking submerged rocks whilst taxiing. In November, 209 Squadron lost its ninth and last Lerwick of 1941 when LZ257 sank at its mooring in a storm. By this time, Catalinas were replacing the unit's Lerwicks and the Saro machines were passed on to No. 4 (Coastal) Operational Training Unit, but two fatal crashes and a further sinking encouraged the decision to use the aircraft to equip a regular Canadian Coastal Command unit, 422 Squadron. The Canadians used all eight remaining Lerwicks to escort Convoy PQ–18, an Arctic convoy of forty Allied freighter ships, from Scotland to the Soviet Union, but no contact with the enemy was made. Two further accidents, one fatal, led to the Lerwick's overdue withdrawal in November 1942. All remaining examples of arguably the worst British aircraft to see combat during the war were scrapped.

During its squadron service, from June 1939 to November 1942, thirty airmen and one civilian technician lost their lives in Lerwick accidents in return for 2,000 lb (907 kg) of bombs dropped on one submarine with no measurable result. The obscure story of the dreadful Lerwick demonstrates how hard-won the successes were that were yet to come.

3. Blackburn Botha (1938)
'Botharation'

Another great Blackburn design, the Botha was damned from a chronic lack of power. Its poor performance meant it was never to enter service in its primary role as a torpedo bomber. Had that been all, it would have been nothing worse than an obscure mediocrity, but Blackburn had also made it extremely difficult to actually see out of the aircraft in any direction except dead ahead. This posed something of an issue for an aircraft now intended for reconnaissance, and the Botha was supplanted by the Anson, which it had been supposed to replace. Passed to training units, the Botha's vicious handling traits conspired with its underpowered nature to produce a fantastic amount of accidents, yet somehow a terrifying 580 were built, and the type soldiered on until 1944.

ABOVE: Terrible, yet relatively plentiful, the weedy Botha was an awful dish served in generous portions.

'Blackburn had also made it extremely difficult to actually see out of the aircraft in any direction except dead ahead'

2. Blackburn Roc (1938)
'Death metal Roc' (or 'A Bad Day at Blackburn Roc')

The Roc was a fairly innocuous flying machine. However, as an example of the wrong concept applied to the wrong airframe to produce a useless combat aircraft, it is hard to beat. The 'turret fighter' that was so inexplicably popular in Britain just before the war was most memorably realised in the Boulton Paul Defiant, an extremely well designed machine (considering its type) that did surprisingly well, given that it had to lug around a draggy, heavy turret to no good purpose. The Roc, by contrast, was lumbered with a massively over-engineered airframe (a legacy of being derived from a dive-bomber), had a less powerful engine and was over 100 mph (161km/h) slower. How an aircraft that could not attain 200 mph (322 km/h) was expected to survive, let alone fight, in 1940 is one of the enduring mysteries of the early war period, as is the fact that its only confirmed kill was a Ju 88, one of the world's fastest bombers.

'The Roc was lumbered with a massively over-engineered airframe (a legacy of being derived from a dive-bomber)'

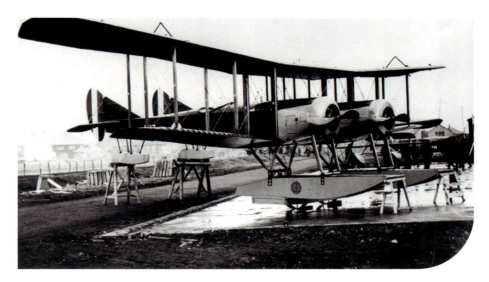

1. Blackburn Twin Blackburn or 'TB' (1915)
'The conjoined flip-flop'

ABOVE: The awful TB was a monster begging to be put out of its misery by its cruel and perverted creator.

BELOW: Speed is life for a combat aircraft, but the Roc rebelled against this holy tenet of survival with an embrace of all that is draggy.

Apparently named after a disease, the TB was a bad aircraft that couldn't perform the one task it was designed for and thus set a precedent for many Blackburns to come. The Twin Blackburn nevertheless saw service for a year or so before it was finally put out of its misery and all nine examples were scrapped. Intended to destroy Zeppelins, the floatplane TB was supposed to climb above Zeppelin height and drop explosive Ranken darts onto any insolent dirigibles foolish enough to approach its precious airspace. Unfortunately, the poor, underpowered Twin Blackburn was unable to drag itself to airship-operating altitude, even after its deadly cargo of explosive darts had been cut by two thirds. Furthermore, the structure, which consisted of nothing more complicated than a couple of B.E.2 fuselages lashed together with a new set of wings and a vast amount of hope triumphing over experience, was not very rigid and the action of warping the wings flexed the poor TB so much it could end up turning in the opposite direction.

The observer sat in one fuselage, the pilot in the other, and communication was impossible except through waving, presumably to prevent either party expressing to the other their true opinion of the designer of this radical machine. As if that weren't enough, the wooden floats were mounted directly below the rotary engines. Rotaries drip out a lot of oil, and as a result the TB's floats would often catch fire. It would be nice to say that, despite all this, the TB inspired the fantastic Twin Mustang, but of course it didn't.

AVIATION MYTHS
YOU SHOULDN'T BELIEVE

Where do the eternal truths we believe come from? Our culture? The Platonic world of forms? Bill Gunston after a pint of bitter? Who knows. Here are four aviation myths we should not believe. I'm locking myself away for a couple of weeks to avoid the inevitable hate mail this will generate.

THE PANAVIA TORNADO F.MK.3 WAS RUBBISH

Though the Tornado bomber remains effective today, the later interceptor variant is long gone. It had a very bad reputation, which in its early days was justified – it had poor agility, poor high-altitude performance and the radar didn't work. But this turd was successfully polished; by the time it was retired in 2011, it was one of the best beyond-visual-range fighters in the world. Former F3 navigator Dave Gledhill noted to Hush-Kit: 'The Stage 3 standard which retired from service in 2011 was light years ahead of that of the F2. At its demise, the F3 was armed with the C-5 standard AMRAAM and ASRAAM missiles, a capable Foxhunter radar that had automatic track-while-scan, joint tactical information distribution system [JTIDS] data link, secure radios, better identification systems and capable electronic warfare equipment including a radar homing and warning receiver, towed radar decoy, flares, chaff [strips of foil or clusters or fine wires ejected to confuse an enemy's radar detection] and a Phimat chaff pod. The situational awareness enjoyed by the crews was arguably better than even the latest-generation American platforms. Regrettably, it still lacked the performance when carrying its role equipment, particularly the 2,250-litre tanks in the upper air, but with improved situation awareness and long-range weapons, the crew should not have been drawn into the visual arena. If it had been employed against an aggressive opponent, the results would undoubtedly have been surprising as it is unwise to underestimate an opponent. The F3 standard at the point of its retirement was one of the most capable fighters in the world and, with further enhancements would have been extremely effective.'

'It had poor agility, poor high-altitude performance and the radar didn't work'

BELOW: A Tornado F.Mk.3 armed with AIM-9L Sidewinders and semi-conformal Skyflash missiles. The Skyflash was nicknamed the 'Sky Slug' by aircrew unimpressed with the missile's pedestrian performance. The later armament of four ASRAAM and AMRAAM was world-class, which contributed to the controversial F.Mk.3 spending its final years as an extremely potent interceptor.

THE CANCELLATION OF THE BAC TSR-2 WAS A TRAGEDY

Declinism is the belief that a nation is heading towards decline, paired with a nostalgic view of the past. The usual stance of the British aviation enthusiast is that the government (typically when Labour) killed wonderful Britain's wonderful aviation industry (sometimes the evils of America or France are also included). The reality was far more complicated (see David Edgerton's excellent *England and the Aeroplane* for more on this). The martyr for this mythology is the TSR-2. Like the F-35, the TSR-2 had a high wing-loading and stacks of leading-edge electronics, and was expected to perform a great many disparate roles. But did the world really need a British vigilante full of prohibitively expensive electronics that would be obsolete as soon as the seventies technology explosion took place? What's more, low-level flying (something the TSR-2 would have excelled at) was the wrong idea, as would prove to be the case in the first Gulf War, when it became obvious that modern air defences make flying at low altitude extremely dangerous.

If it had gone into service, nobody would have bought it – other than perhaps Saudi Arabia (who'll buy anything) and Australia. There would have been no vacuum for the Tornado to fill, and the Tornado started the culture of high-end internationally collaborative combat aircraft that led directly to today's Typhoon. In the absence of either Tornados or Typhoons, the most likely outcome is that Britain would have ended up licence-producing F-15s. These F-15s would actually have been very effective – and a great deal cheaper than the pan-European projects – so maybe the TSR-2 would have been a good idea after all.

THE LOCKHEED F-104G STARFIGHTER WAS TERRIBLE

In an av-geek parallel to the internet's Godwin's law, as online discussion of the German F-104 Starfighter grows longer, the probability of a mention of its allegedly dismal attrition record, or of 'W****maker', approaches 1. A total of 292 Lockheed F-104s were lost in German military service; a catastrophe by twenty-first-century standards. In fact, Starfighter attrition was an improvement over its predecessor in Luftwaffe service, the RF/F-84F. Proportionally, it suffered fewer losses than the RAF's Lightning, that perennial 'pilot's aircraft' (which aircraft isn't?). Long before the Tornado was drafted, the F-104G was blazing a trail across inclement European skies as the first true multirole combat aircraft of the jet age. In Luftwaffe service, the Starfighter was admittedly limited in its roles of interception and reconnaissance, but as a low-level nuclear-strike fighter it provided teeth to back up NATO's rhetoric into the early 1980s. Substitute the additional fuel pack used in the strike role for the M61 Vulcan cannon (which found its first application on the F-104) and hang as much conventional ordnance as that famous tiny wing would permit, and the Starfighter was equally useful in the conventional attack role. The German Navy may have wanted the Phantom or Buccaneer, but they showed just what Kelly Johnson's design could do low over the chilly Baltic, toting anti-ship missiles or running the important 'Baltic Express' reconnaissance mission.

BELOW: Thanks to heavily politicised and deeply nostalgic articles, the ghost of the TSR-2 is never allowed to be at peace, and instead haunts the stinky old mansion that is the British psyche.

THE ENGLISH ELECTRIC LIGHTNING
WAS AN EXCELLENT FIGHTER

We all love the Lightning, but we're often blind to its terrible limitations. An extremely high price was paid for its ultra-high speed and climb-rate: and that price was combat effectiveness. If you are going to have an aircraft with such a pitiful endurance, at least make sure it has a sporting chance of killing the bombers it is sent to destroy. The Lightning's piss-poor radar and two extremely limited missiles (which were short-range with a seeker head that was archaic for much of its service life) meant there was very little margin for mistakes, or bad luck, in an actual interception. By comparison, the single-engined Swedish Saab Draken had half as many Avon engines as the Lightning yet still had a Mach 2 top speed, superb weapon systems and a range more than twice the Lightning's (it was also cheaper to build and buy, better armed and far easier to maintain). The French Dassault Mirage III was another more sensible option.

There was a brief moment in the early sixties when the Lightning was the best, but failure to upgrade it made it one of the worst fighters at the time of its retirement. Scandalously, the Lightning entered the 1980s with no beyond-visual-range weapons (something carried by the Soviet escort fighters it was expected to face) and with prehistoric weapons, sensors and avionics. Still, it wasn't the most obsolete part of Britain's air defence at the time; it was supported by the Shackleton airborne early warning aircraft, which was essentially a Second World War bomber with Second World War radar.

'There was a brief moment in the early sixties when the Lightning was the best, but failure to upgrade it made it one of the worst fighters at the time of its retirement'

SOVIET
FREAK SHOW

Most aircraft are peacock-like prima donnas eager to flaunt their form to a gawping public. Afterburners aflaming as conspicuously as a crotch in tight sportswear, these happy exhibitionists revel in the presence of besandwiched gawpers at the perimeter fence. However, a miserable few seek to hide away their deformed bodies in the shadows. These Phantoms of the Aero-Opera lurk in darkened hangars waiting for the merciful kiss of the acetylene torch. No, we're not talking about the output of Blackburn Aircraft Ltd, but Soviet-era electronic warfare aircraft. So, let's make like Victorians at a freak show and observe some of these crimes against nature.

Tupolev Tu-16 (1952)
'Badger-baiter'

The Tu-16 started life as a moderately attractive aircraft – indeed, the latest Chinese-built Xian H-6N incarnation could even be called handsome. As such, it was easy prey for Soviet aviation's Island of Dr Moreau, with its voracious appetite for disfiguring beautiful machines, where a seemingly endless series of variants were produced, presumably in an attempt to make NATO pilots' aircraft identification quizzes impossible. These included the Badger-D, which looked like a Badger-C but didn't carry missiles; the Badger-E, which appears to have done exactly the same thing but with the full glass nose to give the navigator a panoramic view; then there was the F, which was like the E but with external ELINT equipment in under-wing pods and, on some airframes, an elongated tail to carry jamming equipment in place of the guns.

There was a brief break for the Badger-G conversion programme to add anti-shipping missiles before introducing the H and J models, which had electronic countermeasure (ECM) equipment for standoff jamming or strike escort, respectively. The J model housed an array of equipment in the bomb bay that, it was eventually discovered, was actually interfering with the rest of the aircraft's systems. It probably wasn't doing the aircrew much good either. None of the work to develop this array of niche variants did anything for their looks. Aerials sprouted like mushrooms in a student bedsit, extra radomes were strewn along the belly, and mysterious pods appeared in unusual locations. Although they'd done their best to disfigure the Tu-16, these Badgers were merely the warm-up act for the atrocities to follow.

'The J model housed an array of equipment in the bomb bay that was actually interfering with the rest of the aircraft's systems'

OPPOSITE: With the Il-20M, a once elegant airframe endures a profusion of protuberances suited to various nefarious acts of electronic warfare. The Il-20 is a military version of the Il-18 airliner, crammed with electronic equipment, an array of external antennae and the guilt of its association with Putin.

ABOVE: The intercepting US Navy aircraft that took this photograph would have been wary of the twin 23-mm cannon located in the tail of the Tu-16.

Yakovlev Yak-28PP (1970)
'Soviet Electro Party'

Looking like something out of *Dan Dare*, the Yak-28 was a swept-wing bomber with two Tumansky turbojets in pods under the wings. The Brewer (as it was known to NATO) featured possibly the best-looking bombardier's station of any aircraft, thanks to a nose cone that appeared to be attached to the rest of the aircraft by glass. For no obvious reason other than aesthetics and laziness, this feature was carried over to the Yak-28PP (for *Postanovschshik Pomekh* – 'Countermeasures Aircraft'), which gives it at least one over the EF-111, its later US equivalent. Following the typical depredations inflicted on aircraft cursed with an electronic warfare role, the Brewer-E developed a rash of antenna and dielectric panels, some being mounted on the outboard side of the engine pods. And because this was the sixties, all this electronic trickery was being developed by valves. For the younger reader, valves were used to amplify electric signals before transistors became a thing. For anyone under thirty, think of something like a pint glass with a plethora of electrical connections filling the mouth and mysterious glowing innards. The downside to this was that valves generate a lot of heat, and to prevent it all cooking off, the bomb-bay door had to be covered in intakes for heat exchangers. Intended to escort squadrons of bombers on raids over Western Europe, the Brewer-E was presumably still able to keep up with them despite the extra drag from these modifications. It was also (at least initially) more capable than the relatively clean Su-24MP 'Fencer-E' that replaced it, confirming the direct correlation between ugliness and effectiveness for Soviet Electronic Warfare aircraft.

> 'The Brewer (as it was known to NATO) featured possibly the best-looking bombardier's station of any aircraft, thanks to a nose cone that appeared to be attached to the rest of the aircraft by glass'

ABOVE: A Royal Danish Air Force F-16 shadows a Russian Il-20. Interception allows for a close inspection of any new bumps and blister fairings that may reveal new sensors.

BELOW: Well over a thousand of the utterly adaptable Yak-28 were built. Despite its wartime glory, Yakovlev was slowly pushed out of the fighter business; today it hangs on with the lead-in fighter trainer Yak-130. This illustration shows a Yak-28PP from the 164 ORAP, based in Brzeg in Poland, tasked with escort jamming.

Ilyushin Il-20 (1968)
'The Magic Coot'

The Ilyushin Il-18 'Coot' was an attractive, medium-range turboprop, in many ways the Soviet equivalent of the Bristol Britannia, just slightly smaller, and with an order of magnitude more of them built. We'll pass briefly over the Il-38 May adaption, where the addition of anti-submarine warfare (ASW) equipment saw the wing moved forwards 10 feet (3 m) to address the shift in the centre of gravity, though the general aesthetics just about survived. Instead, let's gawp in horror at the Il-20 Coot variant, which first flew in 1968. This came in two versions, but the Coot-A best perfected the art of ruining a perfectly good aircraft and it's obvious the Coot-B was just a contractually obligated follow-up and the designers' hearts weren't in it. The Coot-A was a Communications and Electronic Signals Intelligence platform, or COMINT, and ELINT in the purposefully baffling lexicon used by practitioners of the dark arts. This required a lot of aerials to listen to all the different frequencies they wanted to cover, and physics being a cruel mistress, these had to be on the outside of the airframe.

Still in service, the fashionable Coot now sports so many domes on its spine that even dromedaries think things have gone too far. Not content with this, the sides of the forward fuselage have sprouted two large boxes, as if the pilots, short of baggage space, had popped to Halfords for a couple of roof boxes and didn't read the fitting instructions. Finally, and slightly bafflingly for a COMINT

aircraft, the belly of the beast features a canoe-like fairing that houses a sideways-looking radar. Coots typically also possess a range of smaller antenna, with no two airframes appearing to be the same, allowing the connoisseur to follow the fortunes of individuals as a naturalist follows individual leopards based on their spots. The IL-20 is, then, very nearly the most disfigured airliner to ever grace the skies, except that the Nimrod AEW3 was briefly a thing.

ROTOR WITCHCRAFT

We now move on to the rotary enclosure. Helicopters are naturally hideous creatures with none of the grace of their fixed-wing compatriots. The Mi-8 Hip family have struggled against this reality since their inception in the early sixties, having a smooth, tadpole-like body that can at least be relied on not to cause fainting fits in young ladies. The Mi-8PP 'Hip-K' version fulfilled a similar role to the Brewer-E in the Soviet order of battle, just without the futuristic aesthetic and moving much more slowly. At first, the looks appear not to have suffered too badly from the transition to the new role – the smooth nose contours remain, with the intakes perched just above the cockpit. But no – what is this? Six, yes six, heat exchangers behind the nose-wheel like remora clutching to a shark. And there's more disfigurement: large boxes stuck on either side of the fuselage below the engine exhausts, containing unknown mysteries. The worst, though, is yet to come: stuck to the aft fuselage are two arrays of six antenna, looking not unlike ceiling fans that have been stuck on a bed frame and then tack-welded to the Hip as it tried to escape. Certainly, this is one of the greatest indignities to face a helicopter, maybe only surpassed by attaching a Searchwater radar to a Sea King with parts from an oil rig. But to what end?

Here we bring in the expertise of Hush-Kit's tame aerospace engineer and technical adviser, Dr Ron 'Sex' Smith, who after some research concluded: 'The main roles are signals intelligence and one form or another of jamming, which could be of communications, weapon sensors or guidance mechanisms. This could include the countering of radars, but might stretch to electro-optic countermeasures. Comms could include preventing radio detonation of IEDs or provoking premature detonation of the same. Signals intelligence can be detecting electronic emissions, locating their sources and classifying them. This may enable command centres to be identified just from signals traffic, without necessarily "reading" any of the traffic. Think about locating terrorist cells from mobile phone traffic.

'An Electronic Warfare helicopter needs to be high enough to achieve intercept lines of sight and will probably operate over "friendly" territory due to the vulnerability of the platform. The Mi-8PP airborne jamming platform initially used the "Polye" ("field") system. From 1980, it was fitted with the new "Akatsiya" system and redesignated the Mi-8PPA, although it retained the NATO designation of Hip-K. It is characterised by the aforementioned "X"-shaped antennas and was built to escort troop-carrying versions of this helicopter and potentially disrupt nearby anti-aircraft gun radars, such as those of the Flakpanzer Gepard.'

The Hip-K, then, was basically intended to operate at a few thousand feet and transmit lots of radio signals to jam enemy radars. Which sounds fine, as long as they don't have any home-on-jam missiles. In which case, to paraphrase Terry Pratchett, it would be like standing on a hilltop in a thunderstorm wearing wet copper armour and shouting: 'All gods are bastards!'

ABOVE: Defiled with a lustily low-brow camo, the MiG-25BM variant took the Foxbat down into the dirty, dangerous world of defence suppression. Armed with Kh-58 missiles, the BM was tasked with destroying enemy air-defence sensors and weapon launch installations. Degrading enemy air defences was a degrading ask for an aircraft used to the refined world of reconnaissance and interception.

THE EMPIRE'S IRONCLAD

INTERVIEW WITH B-52 PILOT KEITH SHIBAN

Since its combat debut in Vietnam, the B-52 Stratofortress has unleashed more destruction than any other aircraft. Pilot Keith Shiban flew the '52 in its nuclear deterrent role and in combat missions over Iraq. He spoke to Hush-Kit about flying and fighting in this menacing enforcer of American foreign policy.

What were your first impressions of the B-52?

I was awed by the size of it. You don't realise just how big it is until you get up close to one. It looks powerful, even sinister, with the dark camouflage and ECM blisters all over it. It's not what I would call a 'pretty' airplane. It's a purpose-built weapon of war and looks the part.

I had just come off instructing in the T-38 and it was like jumping out of a Corvette into an eighteen-wheeler. I didn't find the B-52 difficult to fly, but I did find it hard to fly well. Nothing happens quickly and there is a lot of inertia to manage. Deflect the yoke and there's a noticeable pause before the plane starts to bank. Centre the yoke and it keeps rolling for a bit. After my first training sortie, I can remember looking back at this huge beast sitting on the tarmac and thinking, *Damn, I landed that?*

Air refuelling was for me the most difficult thing to learn. As an aircraft commander, that's where you make your money. If you can't get the gas from the tanker, you can't do the mission. It wasn't until my seventh or eighth training sortie that I was actually able to stay hooked up to the tanker. The short-tail B-52s (G and H models) have a bit of a Dutch roll to them – a potentially dangerous out-of-phase combination of tail-wagging and rocking from side-to-side. It's not really noticeable until you get right behind the tanker. You have to constantly work the yoke just to keep the wings level during air refuelling. Once you finally get your muscle-memory programmed, it becomes second nature, but it took a while to figure it out. Taking on 100,000 lb of gas meant being on the end of that boom for 20 minutes or so. Afterwards, I'd feel like I'd been doing a workout out at the gym.

The other big adjustment was handling that big a crew. The B-52 is very much a navigator's airplane. I used to joke about me just being the voice-activated autopilot for the navigators. In training, I was taught that the aircraft commander's job was to 'fly the plane and make decisions'. I had to constantly process inputs from the other crew positions and decide how to react.

The offence team might be telling me to go one way to get to the target but the defence team might be telling me not to go that way because there's a threat over there. You lived or died as a crew. Even if the pilot is Chuck Yeager (and I'm not), it won't do much if the Radar Navigator can't hit the target or the Electronic Warfare officer lets you get shot down on the way there. It was a team effort all the way. The aircraft commander tends to get all the credit but I was only as good as the rest of my crew. Fortunately I had a very good crew.

How do B-52 crews view *Dr Strangelove*? Was it realistic?

Dr Strangelove was a staple on alert. I've seen it enough times to have the script memorised. Kubrick got an awful lot right with that movie, especially when you consider that the Air Force was very secretive about the B-52 at the time. My main critique would be that the final bomb run seems to take up the last third of the movie, when in reality a bomb run doesn't take nearly that long from initial point to release. I think they tried about eight

ABOVE: The absurdity and enormous peril of the existence of the nuclear-armed bomber force was brilliantly satirised in the 1964 film *Dr Strangelove, Or How I Learned to Stop Worrying and Love the Bomb*. The story concerns an unhinged general who orders an unprovoked nuclear strike on the Soviet Union. Featuring Stratofortresses, the film was popular with B-52 aircrew.

different means of getting those bomb-doors open. In reality, there was a manual release cable that the navigators could pull to unlatch the doors. But hey, it's a movie – they have to make it dramatic. It's still probably my favourite movie of all time.

Talk us through your first Gulf War mission

My crew was deployed in August of 1990 to Diego Garcia, to be part of the 4300 Provisional Bomb Wing. I can remember getting the phone call early on a Sunday morning: 'Be here in four hours with your bags packed. You're going away indefinitely.' After seven months of living on a tiny atoll in the Indian Ocean, the part of me that wasn't scared shitless was ready to just get the whole mess over with so I could go home. It was around five p.m. when we got notified. I know this because the chow hall opened at five and I was getting ready to go and eat. Someone banged on the door to the room four of us shared, and said, 'You're going.' I forget how much time we had to get ready but I know at the dining hall my stomach twisted itself into a knot so all I all managed was a bit of salad and some sips of ice tea.

At the appointed time we were loaded onto a bus and driven down to the airfield. The security police gave us an escort with lights and sirens going, which I thought was pretty cool. We had been previously briefed on what our Night One target would be. We would be hitting one of the Iraqi forward-deployment airfields. There were five of these roughly 25 miles from the border with Saudi Arabia. Three B-52s were tasked against each airfield plus we had a number of airborne spares in case one of the jets broke on

the way there. We had done a few rehearsals against some islands out in the Indian Ocean so were pretty confident in our ability to do it. Our pep talk from the commander was basically: 'Don't run into the ground and do their job for them.' Good advice, actually.

I don't recall exactly when we launched, but it was getting late in the day by the time we actually got out to the aircraft. There was a scheduled time for engine start, taxi and take-off for each aircraft and we launched some twenty bombers and tankers completely by timing, without a single radio call being made.

A fully loaded B-52G is a sluggish beast and needs a lot of runway to get airborne. The runway at Diego was relatively short by Strategic Air Command standards – only 10,000 feet if I recall. We used up most of it by the time we lifted off. There was a nasty line of heavy rain showers hitting the area right around then and we flew through some of it. I can recall taking a pretty good beating going through the weather. A short while after we got levelled off, we did our first air-refuelling. There were normally two air-refuellings scheduled on the way up to the Saudi peninsula. The G-model was a bit underpowered and the extra drag of having bombs on the wing pylons made it worse. Sometimes I would have the throttles to the firewall just trying to stay on the boom.

A good tanker crew could make you look good back there. If they were jinking around a lot, trying to stay in formation, it could make your job a lot tougher. If their autopilot wasn't working, it was even tougher than that. Our bow wave would actually move the tanker around. If either one of us wasn't smooth on the controls, it could cause a chain reaction.

Somewhere on the way up to Saudi Arabia we took time to don our survival gear and sidearms. We had flak vests, as I recall, but I think we placed them strategically around the cockpit rather than wearing them. We figured that anything likely to hit us would come up through the floor.

You probably want to know what I was feeling at the time. I am not a particularly brave individual. I was always pretty scared the days before a mission. Once I got in the jet, I was fine. That was my comfort zone. No more worrying about if it's going to happen; it's happening now. Just do your job.

By this time I was very confident in our ability as a crew to do the mission. We did a lot of training in the six months prior and I knew I could fly the jet to its limits. Knowing that you're probably going to get shot at in a few months gives you added incentive to train hard. It was dark by the time we got up over Saudi Arabia. The sky was filled with the lights of aircraft massing for the attack. I can remember commenting on it, right before I fell asleep. Now, I'd to say that I'm such a steely-eyed warrior that I was able to sleep on the way into combat, but I think I was just exhausted at that point. I had been up most of the day and, combined with the stress, I think I just shut down. Next thing I knew, my co-pilot was waking me up and telling me we needed to get ready for low-level. This involved taping over all the lights in the cockpit with electrical tape and taping green chemical light-sticks under the dash to use as NVG [night vision goggle] lighting. Very high tech. Back then, we had red cockpit lighting that would wash out the night-vision goggles. The goggles were not our primary method of flying low-level but they were an addition to the terrain-avoidance radar and the FLIR that were built into the aircraft. The goggles clipped to our helmet visor and had a battery pack mounted to the back of the helmet with Velcro. The whole assembly was heavy and would snap your neck in an ejection, so you had to remember to take it off before punching out.

Our formation at high altitude was two miles in trail, with each aircraft stacked 500 feet above the one in front of it. As we dropped down to low altitude, we went into what was called a 'stream' – a bomber stream was normally spaced about a minute apart, roughly six miles at the speeds we flew low levels at. We dropped down low, well inside Saudi airspace so we wouldn't get picked up by the Iraqi radars. Our tactics at that time were to avoid known threats. No sense tangling with an SAM [surface-to-air missile] site if you can just go around it. Of course, it's the one you don't know about that worries you.

We were running between 300 and 500 knots on our way to the target. I remember it was pitch black that night and the NVGs

weren't really doing much for me as they need at least some ambient light to work. They were picking up all the anti-aircraft fire, however, and probably making it look closer than it actually was. It looked to me like they were just trying to fill the air with lead and hoping somebody flew through it. I can remember seeing a ZSU-23 spitting out tracers like a fire hose. Fortunately, it wasn't near us because one of those could ruin your day. I saw a lot of heavy stuff – 57 mm and larger. I didn't worry as much about those so much as they had a very low probability of actually hitting something. I occupied myself with calling out what I was seeing to the crew and pointing out that it was either of range or not aiming at us. It's hard to tell what you're seeing at night. Was that light that I just saw a missile or just a truck headlight?

The actual bomb-run was planned as a multiple axis of attack. The three bombers in our cell would come at it from three different directions to confuse the defences. Sixty seconds was normally the spacing between aircraft but in this case we were compressing it to 45 seconds. The idea was to minimise our time over the target. Most critically, we would have to make our time-over-target with zero-second tolerance or our bombs might frag the next guy over the target. The plan allowed no room for error.

My aircraft was loaded with fifty-one cluster bombs that were filled with mines. The other two aircraft had British runway-cratering bombs that we called 'UK 1,000s'. The bombs would crater the runways and taxiways while the mines would make life difficult for anyone trying to repair them. The bombs also had a variable time delay so some of them would dig a hole and then blow up as much as a day later.

To release the cluster bombs, we would have to climb up to

ABOVE: Aerial refuelling extends the already enormous range of the B-52. In 1960, the B-52 Stratofortress bomber set the record for the longest non-stop flight – 10,000 miles – without refuelling.

ABOVE: A B-52 in formation with F-16, F-2 and EA-6B aircraft, showing off the impressive size of the eight-engined bomber. At maximum take-off weight the B-52 approaches half a million pounds, which is more than the weight of 128 adult hippos.

1,000 feet going across the target. That is not a good altitude; you either want to be really low or really high. The other two jets were able to drop from 500 feet. We got the first run over the target so that we might at least have surprise on our side.

The bomb run itself was uneventful except for not being able to see anything. But as soon as we started releasing, things got interesting. In my NVGs I saw *Flash! Flash! Flash! Flash!* and I thought, *Oh crap, they're shooting at us and I can't do a damn thing about it until we get the bombs away.* As soon as the bombs were gone, I went into an aggressive gun-jink manoeuvre. This involved rapidly throwing the plane around in multiple directions. At the same time, I pointed the nose back at the ground. We started picking up speed fast. Our limiting airspeed was 390 knots indicated, and I'm sure I saw 430 on the gauge. At this point the plane was wanting to 'Mach tuck'. The faster we went the more the nose wanted to go down. I had to run the trim nose up quite a bit to counteract that. Meanwhile, we're jinking around low to ground at night, probably being a bigger threat to ourselves than anything

the enemy might be doing.

In 20/20 hindsight, we probably weren't getting shot at but I didn't realise it at the time. What I probably saw that night was the charges from our own cluster bombs opening. The interval between flashes was just about right for it to look like a 37-mm anti-aircraft gun. In all the excitement, we turned the wrong way coming off target and ended up doing a 270-degree turn to get back on course. Meanwhile, the other two bombers did their thing, followed up by a flight of F-15Es, who took out the hardened shelters.

After that, I was hyped up all the way to the Saudi border. The plan was for us to land at Jeddah International Airport in Saudi. I think we had to go around at least twice because the traffic pattern was so busy. Once we finally got on the ground, some guys in silver hazmat suits checked the outside of our plane for contamination (chemicals). Then the maintenance guys checked us for battle damage and didn't find any. Finally we got to park the jet. I can remember sitting on the ramp at Jeddah for a very long time waiting for someone to come get us. We didn't really care; we were just happy to have accomplished the mission and still be alive.

What do most people get wrong about the B-52?

Most people assume that something as large as a B-52 must be roomy on the inside. In reality it's quite cramped in there. Most of the available space is taken up either by fuel tanks, bombs or electronics. The only place you can even stand up straight is the ladder between the upper and lower compartments. Unlike an airliner, it's also extremely noisy. We had to wear headsets or helmets all the time to protect our hearing. Talking cross-cockpit like we do in an airliner was impossible. Everything had to be said over the intercom. Not a comfortable place to spend 12–16 hours. Even training missions would leave you completely drained physically. SAC [Strategic Air Command] liked to say, 'You've got to be tough to fly the heavies.'

What is something less well-known about the B-52?

I don't think our role in the Gulf War was ever well publicised, especially the low-level strikes that were carried out on the first three nights.

During the Cold War, did members of the B-52 aircrew community feel confident that they would survive an attack on the USSR?

That's the big question, isn't it? Fortunately, we never had to find out. Soviet air defences were quite formidable. Our electronic-countermeasures package in the G-model wasn't as good as what the H-model has. There were some newer Soviet missiles, like the SA-10 (S-300) that we simply would not have wanted to meet. We also feared running into a MiG-31 long before we even got to Soviet territory. But you have to realise that by the time we got there, both sides would likely have been lobbing ICBMs at each other for 8 hours. There may not have been much left of their air defences to worry about.

You stood on nuclear alert. How does one reconcile personal ethics with the knowledge one may have to carry out a nuclear attack?

We were so well trained that we'd have probably been halfway to our targets by the time we even thought about what we were doing. We used to joke about turning south and making Jamaica the next nuclear power if the balloon went up but that was just a joke. Most of us didn't think we'd have to do it. The whole reason

SAC existed was to prevent a war with the Soviets. If things had gotten that bad, we'd have probably been dodging nuclear explosions on our way out of US airspace. The instinct would have been to hit them back with everything we had at that point. Still, it was sobering to sign for an alert aircraft with sixteen nuclear weapons on it – quite a lot of responsibility for a twenty-seven-year-old aircraft commander.

Why do you think the B-52 has stayed in service for so long?

In some ways it's such a generic aircraft that it can be adapted to different missions. It can carry a lot of ordnance a long way and it can loiter for a long time. One thing people don't always think about is it has a tremendous amount of electrical power from its four generators. That allows them to keep stuffing new electronics into it.

'We used to joke about turning south and making Jamaica the next nuclear power if the balloon went up but that was just a joke'

Did you ever fly at low altitude in a B-52?

Low-level was our bread and butter in the B-52 community at that time, in the late 1980s. We were still training to penetrate Soviet air defences. In the daytime it was a lot of fun, at least for the pilot – I don't know how the other crew positions managed to sit through it, sitting in the dark while getting bounced around on a hot day was a recipe for airsickness; B-52 navigators are a very dedicated bunch. The downward-firing ejection seats the navigators rode in couldn't have inspired much confidence, either. At night it was very challenging. Our systems were good down to 200 feet over flat terrain and I think 300 or 400 feet in mountainous terrain. Keep in mind that our wingspan was almost 200 feet. A night low-level required a tremendous team effort, especially between the pilots and navs. It was all hand-flown in the B-52. Unlike the B-1 and F-111, we only had terrain-avoidance radar; it wasn't coupled to the autopilot – so, imagine you're bopping along at 360 knots through the mountains in the middle of the night. The SAC tactics people interviewed a Soviet MiG-29 pilot who had defected. They asked him, 'Do you think you could intercept a B-52 flying 300 feet at night in terrain?' He told them: 'No fucking way.'

ARTISTS' IMPRESSIONS

Whether it was the Pentagon scaremongering with images of new Soviet developments or manufacturers whetting the appetite of future customers, our first peep at a new aircraft has often been from a tantalising official illustration. Sadly, most of these aircraft would never really fly, and their exposure was limited to one exciting image. The golden age of these was 1950–1990, when a Boy's Own world of blazing afterburners, exploding enemy tanks and hypersonic airliners was given life in the gaudy, often naive (but always exciting) paintings and drawings happily described as 'artists' impressions'.

OPPOSITE: This dramatic painting was released by Lockheed in the 1980s to promote their efforts in the Advanced Tactical Fighter (ATF) programme.

ABOVE: This McDonnell Douglas ATF proposal was among the lighter and less ambitious offered. It emphasised manoeuvrability and low development costs over low radar cross section.

LEFT: This Lockheed manned fighter concept is built for speed! Flight above Mach 5 comes courtesy of four methane-fuelled turboramjets.

TOP LEFT: Early DoD artwork of the MiG-29 was reasonably accurate, aside from the F-16-shaped front fuselage and peculiar cockpit canopy.

TOP RIGHT: The Model 225A was McDonnell Douglas's entry for the US Navy's VFX competition in 1968, which eventually led to the Grumman F-14 Tomcat. The 225A concept featured canard foreplanes and a variable-geometry 'swing wing'.

RIGHT: The Advanced Deck Launched Interceptor was probably designed around Marquardt's Supercharged Ejector Ramjet (SERJ). Its role was to launch, scream out to a few hundred miles and destroy incoming Tu-22s before they could fire missiles.

ABOVE: The Lun-class ekranoplan (Project 903) was a ground effect vehicle. Longer than a Boeing 747 and with an all-out weight approaching 380,000 kg, it vies with the XB-70 for the title of most awe-inspiring military vehicle ever flown.

LEFT: We banned rotorcraft from this book, but somehow this helicopter has been smuggled in. This was a Boeing–Sikorsky proposal for the Light Helicopter Experimental Programme, a 1980s United States Army helicopter procurement effort to replace the AH-1 Cobra and OH-58 Kiowa.

RIGHT: The 200 was a 1973 naval carrier fighter proposal that was planned in both VTOL and CTOL variants. The VTOL variant included a jet exhaust pipe tiltable by 90 degrees (the nozzle used a three-bearing system) and additional lift engines.

BELOW LEFT: This 1980s Grumman Advanced Tactical Fighter proposal features a 'piggy-back' dorsal semi-conformal carriage for what are either guided weapons or unmanned air vehicles.

BELOW RIGHT: In 1957, Convair proposed a novel system to USAF utilising the supersonic B-58 Hustler as a 'mothership' carrier for a new high-speed parasite aircraft. The project was named 'Super Hustler'.

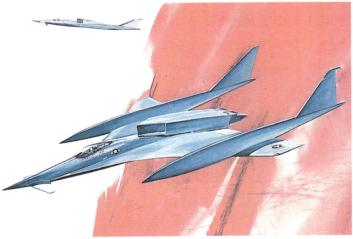

ABOVE: Grumman's STOL Fighter concept, derived from the earlier Advanced Technologies for Tactical Aircraft studies. This impressive aircraft would have relied on flapped canards and vectored thrust for a quick departure from short or damaged runways.

BOTTOM LEFT: The Grumman ATS was an intriguing design from 1976: the placement of the main fuel tanks in mid-wing booms allowed for a neat short fuselage.

BOTTOM RIGHT: Britain loves the idea of vertical, or very short, take-off fighters. This BAe study was the P112.

THE ROARING DOUBLE-DECKER

INTERVIEW WITH ENGLISH ELECTRIC LIGHTNING PILOT IAN BLACK

English Electric Lightning – three words that sit so beautifully together. The English Electric Lightning was the most exciting jet fighter ever created, the charged air of English skies ripped apart by its riveted lunacy. The Lightning was quite mad: a greedy machine set on eating fuel and turning it into speed. Unlike anything else, its two Avon engines were stacked one on top of the other, making it stand monstrously tall on the ground, and when it entered service, in 1959, it was the most formidable fighter in the world; for twenty-nine years it thundered over British skies as a brutish deterrent to would-be attackers. Ian Black flew this overpowered monster for the Royal Air Force in the final years of the Cold War. Here he shares the secrets of flying and fighting in what could remain Britain's final jet fighter.

What does a Lightning cockpit smell like?

Winner of the most unusual question! Are you going to market this as an aftershave? 'Smell like a fighter pilot. . .' If you blindfolded me, put me in a cockpit and said 'inhale', I could probably identify 'fighter cockpit'. . . 'Lightning'. . . 'Mk.6'. Given they were twenty-five years old, I'd say a heady mixture (to use advertising speak) of sweat, old leather, fuel and warm Bakelite. Then add a hint of rubber from our exposure suits and a touch of burned cordite after gun-firing. A flick of a switch and a blast of 100 per-cent-pure oxygen masked any aromas and gave you a breath of fresh air, as they say.

If the Third World War had started while you were on the Lightning force, what would you have expected from Day 1?

That's a good question, and I will try to answer honestly and, more importantly, with a sense of realism. My thoughts are that Day 1, or pre-Day 1, we would have seen an escalation of tensions – so, lots of probing flights into UK air space by 'Bear' bombers on ELINT recce missions. And now the realism part: initially no Soviet fighters or bombers would have been able to reach UK. The likes of MiG-23/27 and MiG-29 just didn't have the range – nor did the MiG-25, for that matter. My guess is that, the first few days, we would have been intercepting long-range bombers whilst avidly watching the TV as hordes of Russian and Eastern bloc fighters pushed west.

We had two jets on 10-minute state on Q [Quick Reaction Alert]. This would have been increased to have 4–6 on 10-minute state. Think about the logistics: twelve aircraft on the squadron and fifteen pilots – we couldn't have kept it up for long.

What were your first impressions of the Lightning?

Very big – it sat high off the ground, unlike the Hawk. It seemed to have myriad switches, all randomly located in the cockpit. It was very cramped when wearing full exposure suit, which we did

for eight to nine months of the year in the UK. It had an extremely eccentric starting system that was a bit like a Jules Verne rocket, and once the engine was turning it was like lighting a firework and you were off on a journey of a short but exciting duration.

What were the Lightning's worst vices?

Lack of fuel was the obvious one. From a handling point of view, it was gloriously over-powered, a feature few aircraft have. With its highly-swept wing and lack of any manoeuvre/combat flaps or slats, the aircraft was often flown in the 'light-heavy buffet', which masked any seat-of-the-pants feeling of an impending stall. It actually had few of the traditional vices but could be a handful on landing with its big fin and drag 'chute, which made the aircraft weathercock on a strong crosswind landing. Tyres were also by necessity very thin to fit into the wing, and high-pressure, so didn't last long.

How good was the radar?

In 1960, when the Lightning entered service, it was probably state-of-the-art, but by 1988 it was positively prehistoric! It was hopeless at low level overland, difficult at low level over the sea. At height, the targets would often be doing in excess of .9 Mach, so the combined speed of fighter and target would be around 20 miles a minute; with a maximum pick-up range on an average target of 18–20 miles, this gave you less than a minute from initial contact to engagement. It also had very limited electronic-countermeasures capability.

Could the Lightning be described as having HOTAS?

That's a very good question. . . but the answer is no. Mainly because the throttles only had the transmit switch and air brakes; on the control column, just the gun camera and autopilot, so no weapons controls on throttles or stick.

How would you rate the Lightning cockpit for ergonomics?
I'd say for the 1950s/1960s, it was a five out of five; you could pretty much get to any switch with your eyes closed and the radar hand-controller was a work of art in itself.

How good were the weapon systems?
Again, the weapons system was state-of-the-art in the 1960s; by 1988 it was prehistoric. The system, a data-link where the ground controllers would perform the intercept with the pilot flying to target hands-off, had potential. The weapons were fine against lumbering Soviet bombers up at altitude, but not great in a high-G combat scenario.

How much faith did you have in the missiles working?
They were old but pretty basic. I guess the success rate might have been 50–60 per cent hitting the target – but I'd be pretty sure they would come off the rail.

'There might be a jet – say, tailcode "DD", for example – that was serviceable for three to five days with no snags, then a few faults would appear and it would go in the hangar – a cursed one'

Shooting down Soviet bombers over the UK: was this discouraged? Was it considered better over the sea?
I don't think it mattered where you shot them down! If they committed a hostile act and the rules of engagement allowed, we would have shot them down.

How much difference was there between individual airframes?
They were definitely all unique – notwithstanding the fact that the RAF had decided to paint the 'Silver Machines' [Lightnings initially served unpainted]. They were initially dark olive-green, then camouflage, then they went Air Defence Barley Grey – and

then they settled on a dark sea green in various guises. So, no two looked the same.

Some were definitely reliable, but it was normally only short-lived. There might be a jet – say, tailcode 'DD', for example – that was serviceable for three to five days with no snags, then a few faults would appear and it would go in the hangar – a cursed one. Others were always unserviceable, and the ground crew knew which jets to take away on detachments that would not let them down and which ones to leave behind.

Best and worst Lightning marks? And why.
Well, I only flew the F.Mks. 3, 5 and 6 – but for me, the 3. It was as capable as a Mk.6 but had half the fuel. That meant you would be back on the ground in 40 minutes and off to do it again!

Fastest and slowest Lightnings?
I gather the F.Mk.1s were fast, but then again they had the smaller, rounded fin that was not really stressed to go Mach 2. For me, a Mk.3 – oh, and a Victor tanker close by! Slowest was the F6 with over-wing tanks on.

Do you think the UK was well defended in the 1980s?
I think the eighties were a pivotal time for the RAF; they definitely went from the old valve era into the digital age. Much changed for the better: GCI [ground-controlled interception], the Tornado F3, Link 16, and in particular the acquisition of the E3 AWACS – magic.

How did Lightnings do against teen-series fighters in dogfight training exercises? What tips would you offer in these situations?
Lightnings fought F-14s, F-15s, F-16s and F-18s. At long ranges, Lightnings would have been shot down with radar-guided missiles – with no radar warning receiver, the Lightning would not have stood a chance. Against the teen series, the Lightning did OK in close-in combat, but the best version for air combat was the F.Mk.3, and that had so little fuel you could really only last for one engagement.

If you're fighting a Phantom in a Lightning, what is the best approach?
Use the vertical – keep the F-4 close and keep it high, where it doesn't perform as well – around 5,000 feet, a clean-wing F-4 (UK) was a close match for a Lightning. If you were fighting an F-4 with AIM-9L, it was a hard match, so keeping it tight and trying to be

inside his minimum range was good. . . and use guns.

How would the Lightning have done against a MiG-23?

Easy. The MiG-23 was pretty awful at a turning fight, but would probably have outrun a Lightning at high-speed at low level.

Which tactics should Lightning pilots use in air combat?

My own tactic was to come to the merge at high speed – say, Mach 1.1–1.2 – then to come back to idle at the cross point to avoid getting shot in the face then start a low-G climbing turn with full reheat, hoping to top out around 40,000 feet (making sure you didn't go into contrails and give your position away). If your opponent didn't climb up with you, it was an easy task to dive down on them (they were often now blind to you) and pick your moment.

What were your most notable flights?

Flying my father, taking a Lightning to Cyprus twice, flying low-level in West Germany from Gütersloh, where Lightnings had been based in the 1970s.

Flying my first Lightning solo was incredible. Imagine watching something you loved for twenty-five years and then actually getting a chance to do it – but in the process you have to learn to be a fighter pilot! So, flying a Lightning solo was pretty special, but taking one across the Med with a tanker was a unique experience – I flew a T.Mk.5 once, to Cyprus and back, and an F.Mk.6 one way. The T.Mk.5 had to be refuelled six times to get there, with the aid of tanker support.

As a child, I had always assumed flying Lightnings at low level in Germany was as good as it got (over the North Sea wasn't nearly as exciting), so given the chance to do a week of just that was too good to be true, especially as we were working with the Harrier force, engaging in air combat when the weather was too bad to fly at low altitude.

Taking my dad flying was a bit nerve-racking – I had 50 hours on type, while he had nearly 2,000. It was fifteen years since he'd last flown Lightnings, and he was regarded as one of the best Lightning pilots there ever was. During the flight he pretty much flew it from start to finish. I'm not sure what was worse – him teaching me to drive, or me taking him in a Lightning!

How well trained were Lightning pilots? Were you given sufficient flying time?

Lightning pilots, along with Harrier pilots, were the best – no contest. We got lots of flying, and we were always on top of our game, from low-level intercepts to high-flying, supersonic targets.

What should I have asked you about the Lightning?

What makes the Lightning unique. It's the only jet fighter with a vertical twin-stack engine-layout. It's all British and did Mach 2.0 It's probably the ultimate fighter in terms of man and machine working as one.

CAPT BRENT "STRETCH" BLACKMER
DCAG

A BRIEF HISTORY OF FIGHTER COCKPITS

Flying twice as fast as a round from an AR15 and capable of pulling G-forces that leave pilots with the same painful lack of mobility as if they weighed an actual ton, a combat aircraft asks a lot of its pilot. Fighting and surviving in such a hostile environment requires lightning-fast assimilation of a mass of information and an equally fast response to it. Not only this, but today most fighters are multirole and tasked with destroying both air and surface targets. This is possible thanks to the wonder of the modern cockpit. We asked former TOPGUN instructor and F-14 radar intercept officer (RIO) Dave 'Bio' Baranek to give Hush-Kit the lowdown on the fighter pilot's 'office'. So, let's slam the canopy shut and take a flight through sixty-five years of design.

Fighter-cockpit design has evolved with two key concepts in mind: give the pilot more information and make that easier to understand. In 1943, a lieutenant in the US Army, Alphonse Chapanis, demonstrated that pilot error could be greatly reduced with more logical and easier-to-identify controls and displays in cockpits. Following the war, the USAAF published a nineteen-volume study of what had been learned from wartime experience. Since then, an increasing awareness of 'human factors' and ergonomics has radically altered cockpits to give pilots every possible advantage in making good decisions in complicated situations. The science of cockpit design took pilots from the confusing mess of many fifties and sixties designs to the large-screen, 'God's eye view' of the most modern cockpits. Sixty-five years seems like a long time ago, but the F-106 Delta Dart that starts my list could be a threat today if still operational; its near-contemporary, the F-4 Phantom, is still in service.

I was a TOPGUN instructor and an F-14 RIO, but for this article I'll move into the front seat and look at instrumentation and controls. This is not an exhaustive survey, but a look at representative types that I have selected. I'll address the earliest version of each type because later developments had more to do with technical advancements than the state of aircraft design. Imagine a Spitfire Mk.24 with a podded radar, helmet-mounted cueing system and ASRAAM – with the controls and displays to support it all – and you get the idea.

'ICS check. . .?' 'Loud and clear.' 'Okay, let's get going.'

CONVAIR F-106A DELTA DART (1956)

As one of the later Century Series aircraft, I will argue that the F-106 was the dawn of the modern fighter. The Delta Dart was publicly described as a development of the F-102, but it was a radical improvement (the long-awaited realisation of the very ambitious '1954 Interceptor' concept). Whereas the F-102 cockpit looks like something out of a hobbyist's basement, the 106's looks like a fairly modern fighter/interceptor, at least before the dawn of glass cockpits. The tape instruments add a modern touch, and the fact that it's single-engine allows the panel to be less cluttered than dual-engine types. Former pilots of the F-106 have described the procedure to select weapons as 'cumbersome' and difficult to accomplish under combat conditions. Such realisations were sweeping the aviation industry and led to modern HOTAS cockpits.

As a teenager, I met a pilot who flew F-106s in the Florida Air National Guard, based in my hometown, and he arranged for me to fly their simulator during one of my visits to watch them fly. I was pretty excited, and to my surprise discovered that I was able to avoid crashing – with a lot of coaching from the simulator control console. The moving map display in front of the control stick was cool; it seemed futuristic in the 1970s.

MCDONNELL DOUGLAS F-4B (1961) AND F-4C PHANTOM II (1963)

In early Phantoms, the pilot instrument panel is similar to the F-106 in level of complexity. With a back-seater to handle the radar, the F-4 didn't need a two-headed stick like the F-106. One element that doesn't show up in the cockpit photos is the relatively poor outside visibility of both of these early aircraft; it just wasn't a priority. But at least the F-4 pilot had a head-up display (HUD), while the F-106 pilot had a large radar scope in front of his face. The Phantom HUD was likely deemed essential to its strike-fighter role.

GRUMMAN F-14A TOMCAT (1970)

In the F-14A Tomcat, I like the arrangement of critical flight instruments in an upper tier, with engine instruments and a situation display below them. The stick and throttle have numerous switches and buttons supporting HOTAS. The forward control panel looks relatively simple compared to the contemporary F-15A, which can be at least partly attributed to the Tomcat having a rear cockpit for armament control switches and other controls. The F-14A pilot's primary tactical display was a repeat of the radar intercept officer's tactical information display (TID), so crew coordination was important. The F-14A HUD was helpful in some situations but most pilots decided it wasn't that good: when it displayed all info, it was cluttered and not what a pilot really wanted, and in the declutter mode it didn't display very much at all. This was finally fixed in the F-14D, which got an improved HUD. The large canopy provided excellent visibility, which was one of many lessons from Vietnam air combat incorporated into the F-14.

GENERAL DYNAMICS F-16A FIGHTING FALCON (1974)

The F-16 sported a relatively uncluttered cockpit for a multirole fighter, which can be attributed to factors such as being single-engine and placing an emphasis on the HUD, as well as the limited air-to-air radar in the A-model. There is also the matter of good design, of course. The monochrome tactical display was low and centred, with primary flight instruments immediately above. Cockpit visibility was outstanding due to the lack of a canopy windscreen bow and high-mounted seat. The side-mounted control stick pioneered in the F-16 has become familiar on other modern fighters and some commercial aircraft.

SUKHOI SU-27 FLANKER B (1977)

The Flanker B's cockpit is roughly similar to the F-14 and Tornado in terms of visual complexity, with one major difference: no video screen in the centre. This lack of a tactical overview display seems to me a reduction in situational awareness, even if the pilot is using a helmet-mounted display (the early Flanker pilots had a rudimentary helmet cueing system rather than a display). Nonetheless, it was equipped with the now-standard HUD and HOTAS, and the high seating position and bubble canopy provided excellent visibility. The cockpit was less cluttered than the MiG-29 (which also first flew in 1977), probably because the bigger size provided more room for displays and controls.

PANAVIA TORNADO F3 (1985)

This is another pilot cockpit that benefits from shifting some controls and switches to the back seat. The F3 instrument panel is uncluttered and features two medium-size video screens, one directly in front of the pilot. HOTAS: check. HUD: check – with extra points for wide angle – and of course there's the wing-sweep controller. The neat and well-organised layout is very appealing. One reason for this is that the gauges are one of only three sizes; in many American fighter cockpits, each instrument seems to have a unique size. The Tornado's is probably one of the best cockpits before 'glass' took over and gave us multi-functional displays (MFDs). The Tornado also has a generous canopy, although it doesn't have the 360-degree view of other fighters.

DASSAULT RAFALE (1991)

It's hard to believe the Rafale has been around for over thirty years since its first flight! The cockpit still looks modern and uncluttered. This is possibly due to the control stick being on the right side instead of central. The throttle has display image controls, ensuring a strong finish in the battle for who has the most HOTAS buttons. The wide-angle HUD, bigger than on previous aircraft, has to be a welcome development for almost any mission. The central screen is a head-level display (HLD), in Dassault terminology: larger than the side screens, it improves the pilot's view of the image from a targeting pod. The Rafale's HLD is also focused at a greater distance than the screen's actual distance from the pilot, which allows the pilot's eye to remain focused at near infinity whether looking through the HUD or at the HLD, instead of changing focus between infinity and 1 metre.

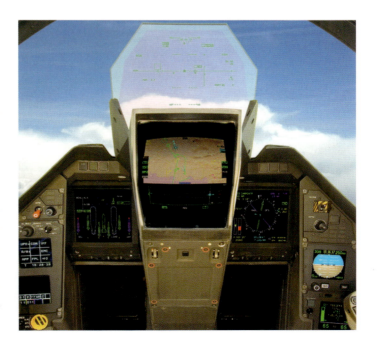

EUROFIGHTER TYPHOON (1994)

The Typhoon is frequently compared with the Rafale (above), but they have different purposes and strengths. To my eye, the Typhoon cockpit doesn't look as sleek as the Rafale's because the Typhoon has more controls. The Typhoon is more spacious, although I must admit the Rafale appears adequate. Like the Rafale, the Typhoon also has a wide-angle HUD. The Typhoon's multiple MFDs and pilot-personalised displays look like a great way to display huge volumes of information very effectively. Like the Rafale, the Typhoon has a voice-input system. I know these things are tested extensively before being fielded, but based on current voice controls I am suspicious of their efficacy. The Typhoon also has the benefit of a mature helmet display/cueing system, something that only arrived a fair bit later in the Rafale community (for some export customers).

BOEING F/A-18E SUPER HORNET (1995)

The Super Hornet cockpit appears similar to the Typhoon – modern and well organised – with some notable exceptions. First, the Super Hornet doesn't have a wide-angle HUD. On the plus side, there are back-up essential instruments tucked into a corner of the panel. Call me old-school, but I like having the back-ups. I also like the glare shields protruding from the top of the SH panel.

LOCKHEED MARTIN F-35 LIGHTNING II (2006)

The biggest attention-grabber in this cockpit is the single large screen, with touch controls so extensive we see relatively few switches and controls elsewhere in the cockpit. The originator of the big screen was Gene Adam, who was at the McDonnell Douglas aircraft company in St Louis. He was predicting big-picture flat-screens in aircraft way back when a TV was the size of a camping rucksack.

I hope this cockpit never fails, because I don't see any back-up instruments. Other attention-grabbers are the side-stick location and the lack of HUD, which is replaced by the pilot's helmet-mounted display (HMD). The F-35 is establishing a new standard for fighter cockpits, with a similar large single display planned for the Gripen NG and Block III Super Hornet upgrade. The designed integration of the large display and the HMD will give F-35 pilots a very high level of situational awareness on any mission. I had a candid discussion with an F-35 pilot who I knew would unload if they had any complaints. But he smiled and said the new jet was 'incredible' and 'awesome'.

And now a final thought, something any aviator can tell you. If you look at these images and think the cockpits look complex, it's only because you don't have experience in that type. The first time I saw the rear cockpit of an F-14, with dozens of panels and controls, I was stunned. But after completing my training and then flying more frequently (I averaged 39 hours a month during my first few months in a fleet squadron, in 1981), I realised I was reaching for switches and adjusting controls almost subconsciously. Training will be the key for pilots to employ these cockpits, no matter the design features or flaws.

F-35 PILOT'S VIEW

Obviously I'm limited in what I'm allowed to tell you about this machine, so I'll stick to what's available in the public domain. First up, there's no HUD as it's all integrated into the helmet. The technology of the helmet is great, but I'd take a HUD any day. It all comes down to physics – you can only shrink things so much before they start to become degraded, and HUDs have bigger optics than helmet – currently, at least.

The side-stick is something I thought would be difficult to convert to, but in all honesty it was a non-event. The rest of the cockpit is beautiful to look at – nothing analogue, all digital with only about ten actual switches in the cockpit. Notice I say beautiful to look at, not necessarily beautiful to interact with! In theory, the all-glass display is great. It's touchscreen, and you can set it up to show pretty much anything you want, in any layout you want. Take a fuel display, for example. You can have it in a large window that shows you everything you could possibly want to know about the aircraft's fuel system – the contents of each tank, which pumps are operating, fuel temperature, centre of gravity, etc. Or you can shrink it into a smaller window that only shows more basic info. Or you don't even display it at all because the function access buttons (FABs) along the top of the display always include a small fuel section, with the essential info visible at all times. That's the beauty of the display – size and customisation.

The drawback is in the complete lack of tactile response. It can be challenging to press the correct 'button' on the display whenever the jet is in motion as it is quite a bumpy ride at times. I currently press the wrong part of the screen about 20 per cent of the time in flight, due to either misidentification or, more commonly, because my finger gets jostled around in turbulence or under G. One of the biggest drawbacks is that you can't brace your hand against anything whilst typing – think how much easier it is to type on a smartphone with your thumbs versus trying to stab at a virtual keyboard on a large tablet with just your index finger.

Voice input is another feature of the jet, but not one I have found to be useful. It may work well on the ground in a test rig, but under G in flight it's not something I have found to work consistently enough to rely on. I haven't met anyone who actually uses it.

Having bashed the interface, the way this jet displays information to you is incredible. The sheer amount of

'Voice input is another feature of the jet, but not one I have found to be useful. It may work well on the ground in a test rig, but under G in flight it's not something I have found to work consistently enough to rely on'

situational awareness I gain from this aircraft and its displays is like nothing I've experienced before. The off-boresight helmet is much more accurate than legacy Joint Helmet-Mounted Cueing Systems (JHMCS) and I find it clearer to read. About the only thing missing from the whole cockpit is the lack of 'feel'. But I still want a wide-angle HUD for flight and fight-critical data. One source I spoke to explained some of the background to its absence. 'The HUD story on the F-35 is a sad one. They wanted the panoramic display really badly – it had been Wright-Patterson's pet rock for years – but in 1995 the display and the HUD both needed to occupy the same space. Hence the HMD having to replace, rather than supplement, the HUD, which caused a lot of problems. But then, by 2010 or so, flat screens and digital HUDs were standard and it was no problem to install both (as on Boeing fighters). Unfortunately, it would have cost the usual squillion dollars to put a HUD on the F-35.'

TAKING THE F-4 PHANTOM II TO WAR

INTERVIEW WITH IRANIAN AIR FORCE PILOT ALIREZA NAMAKI

The scale and brutality of the Iran–Iraq air war of the 1980s was astonishing. We spoke to Brig. General (Rtd) Alireza Namaki, a former squadron and wing commander at Bushehr tactical fighter base, who commanded an F-4 fighter wing and survived numerous combat missions against Iraqi targets. He shared his insights into the potency of the Phantom, the raw drama of ground-attack sorties, frustration with bad leadership and the appalling horror of a one-man mission of revenge that will forever haunt him.

Three words to describe the F-4 Phantom?

Allow me to say four words. On top of being the furthest-flying, highest and fastest fighter plane of its time, it was also reliably a 'pinpoint' striker. Plus, it could carry more than 10,000 lb of ordnance, and could be armed with radar-guided and heat-seeking missiles. Its later version, the E-model, was armed with an internal gun, too. It could fly up to 50,000 feet at sustained speeds unrivalled by its contemporaries. It could strike any targets anywhere.

Its best traits?

In its own era, it was the best. Its bombing computer, armament, fuel capacity and its ability to be refuelled mid-air were unique for its time. It was technologically ahead of its rivals like the MiG-21 and MiG-23, and among its Western colleagues it was top notch, certainly better than the A-4 Skyhawk or early versions of the Panavia Tornado. But since you asked, the aircraft's biggest and most useful trait was its forgiveness; it could tolerate a pilot's mishandling and mistreatment of the airframe better than almost all other jets. It was sturdy, too – it could take a beating. Many pilots survived their ejection, bad landings and combat solely due to the F-4 being a superb aircraft. It is now clear that the F-4 was the ultimate 1970s multirole war machine, and it could also be employed in a strategic role for smaller nations like Iran. A case in point is the Iranian air force's strategic attack on the Osirak nuclear reactor in late September 1980 (on the seventh day of the war, in a two-ship strike mission led by Major H. Ghahestani), which forced French engineers and support personnel to leave Iraq the following day. I am prepared to argue that Israel's attack on Osirak later on was largely symbolic, since we had already inflicted severe damage to the facilities. Another strategic strike came during our several missions against the Salman Pak nuclear [or more probably chemical and biological warfare] research facilities south of Baghdad throughout the first year of the war.

Tell us something we don't know

The late Shah's air force acquired the F-4D/E in large numbers to satisfy a strategic need at the time. Our neighbour to the north, the Soviet Union, was a menace. Our neighbour to the west, Iraq, was a threat. The Phantom was purchased to deal with the threats of its time. No presidential palace, no oil facility, no air base was safe from our reach. It could fulfil a strategic role for our air force as well as a tactical role.

Another interesting point is the late Shah's desire to buy the F-111 instead of the F-4D/E. It is why I use the word 'strategic' for the role F-4 was to play for us.

How satisfied were you with F-4's weapons and performance?

You are asking me how satisfied I was? I am an F-4 pilot. I was and am delighted by the F-4's performance, including its armaments. In the 1980s, a few F-4E Phantom fighters, each armed with half a dozen AGM-65 Maverick missiles, could destroy any ship at will. This was proven in late November 1980 during Operation Morvarid, in which we essentially removed Iraq's naval capacity from being an effective factor in waging war in the northern Persian Gulf.

How was life during the war in and out of squadrons? What lows and highs did you personally experience?

The answer to this question can be as long and extensive as the eight years of the war itself. On the first day of the war, I had to kiss my family and children goodbye. As we lived in a war zone, at Bushehr air base, they had to be evacuated to a safer city beyond the reach of Iraqi fighter-bombers. I managed to go home once, after thirty days straight. The house was extensively damaged, and its windows were completely shattered due to Iraqi bombs going off nearby. I felt I was taking revenge for my own destroyed house

ABOVE: With a strategic reach and the ability to take a beating and still get home, the F-4 Phantom II was beloved by the Iranian aircrew who took it to war. US-made Maverick air-to-ground missiles proved devastatingly effective.

'Mankind created war, just as it invented lying and dishonesty. And I abhor what I did'

as I led a four-ship formation to bomb Az-Zubayr oil field west of Basra. As a war-fighter, I was truly hurt whenever the Iraqis would attack our population centres and we were absolutely forbidden by our own government to retaliate in kind.

In one such attack, an Iraqi jet struck a girl's middle school near the city of Abadan, resulting in the death of more than twenty-three students and a young teacher. This event caused me a lot of emotional pain. I think I had found my own reason, or excuse, for a personal vendetta. This terrible incident was seared in my mind until the day I was tasked to lead a three-ship sortie call sign Houman to attack the town of Khor Al-Zubair's steel and iron plant 40 km south of Basra. Each aircraft was armed with six BL755 bombs. These are cluster bombs designed to destroy tanks and armoured troop-carriers. That day I got to take my revenge and wage my own personal war. I had decided to save one of these bombs to drop on Basra on our way back to Bushehr air base. My reasoning was to give the Iraqis a taste of their own medicine. Choosing a north-to-south heading, I released the remaining bomb on what appeared to be an empty street, dove to 20 feet in afterburner while dodging a hail of AAA arcing over my canopy. That very night, Radio Baghdad reported that upwards of forty Iraqi citizens had been killed and wounded in a bombing raid. Our wing commander summoned me and questioned me. The air force headquarters was desperate to find the perpetrator. But I denied it and they eventually let go of it. I am now a retired warrior and I absolutely regret this incident. It is apparent that what I had

done was, and is, against the accepted norms of humanity and was against international law. Mankind created war, just as it invented lying and dishonesty. And I abhor what I did. I am not proud of it.

Such harrowing combat tales are aplenty. The regular Iranian armed forces did not target Iraqi civilians. It is important to add that, in the later stages of the war, in what came to be known as the War of the Cities missile strikes against Baghdad, our regular forces did not conduct such attacks.

To be honest, my flying career is now defined by the gruelling years of the Iran–Iraq war. Another event stands out. By 1987, I was a fighter wing commander [in the Iranian air force, a wing commander is also the base commander] at Bushehr air base in the latter stages of the war. Our wing was instructed by our higher headquarters to strike three oil tankers carrying Iraqi oil or heading to Iraqi oil terminals per week. This is during the early stages of the infamous Tanker War. It was the weekend, and our quota of three had not been achieved. We had managed to hit two ships that week. Our F-4 jets had found a third vessel to attack, but Saudi Arabian F-15 fighters had arrived and prevented our side from performing a successful strike. This third vessel's captain had now decided to deviate and head for Saudi's Ras Tanura port. Hitting a vessel in a neutral country's waters was akin to declaring war on that neutral state, and as such it was not advised. Time was running out, and this vessel had to be hit before it took refuge in Saudi Arabian territorial waters. I decided to fly this special mission myself, as I did not want to endanger the lives of my younger pilots. Flying as low as I could to close the distance, I popped up around 8 miles out and fired two AGM-65 Maverick missiles at it, dove back down and flew straight to Bushehr AB as low as was feasible. This definitely put an end to the non-stop messages my office was receiving from Tehran on the need to strike three ships a week.

BELOW: Resplendent in 'shark's teeth' and three-tone desert camouflage on its upper surface, this Islamic Republic of Iran Air Force F-4 Phantom II is armed with AGM-65 Mavericks and a 20-mm Vulcan rotary cannon.

How valuable was the Phantom for Iran, both in terms of how it was perceived and its combat capability?

In the early to mid-1970s, Iraq's leadership grew increasingly hostile towards Iran, so our war planners in the air force began concentrating on countering this probable threat. Plans were made to attack all Iraqi airfields on the first day of hostilities to deny the Iraqi Air Force a chance to use them for further aggression. Our pre-1979 contingency plans had us bomb each Iraqi airfield with roughly fifty aircraft. This objective required more or less 300 strike aircraft to fly in a single day to secure air supremacy for the ground forces to advance inside Iraqi territory.

It is widely accepted that our air force performed a deterrent role before its personnel were decimated by the Islamic revolution and the ensuing purges. Psychologically speaking, Iran's neighbours were informed of our capabilities and were aware that any air or land strike against Iran had to be foolish since it would be responded to with overwhelming force. Combat-wise, this aircraft was unique in what it brought to the table. As I've said, this aircraft flew the furthest and highest among its contemporaries. I remember vividly that, back in 1969, during a bilateral training exercise with the Pakistan Air Force (I had just finished my initial F-4 combat training) we intercepted a Pakistani Canberra bomber flying at 45,000 feet. Up to that point we had no fighter aircraft that could do these types of missions. The aircraft made an invaluable combat contribution as a whole.

The Islamic Republic of Iran Air Force lost nearly 50 per cent of its F-4 Phantom II aircraft during the war and this is most hurtful to me. Some of these losses were unnecessary and could have been avoided had our side employed people with a degree of professional knowledge instead of employing religious zealots who had weaselled their way to the top and did not know anything of tactics or warfighting.

Anything else you'd like to share with readers?

Maybe one last thing. In one particular 'max range, max performance' strike mission early in the war, we flew a handful of F-4s to attack a commercial transit point, also known as Arar border-crossing on the Iraqi–Saudi border. We knew this border-crossing to be the place from which Iraq was importing aviation fuel, lubricants and diesel. The other port of entry, Safwan at the Iraq–Kuwait border, was closed as a result of weeks of air bombardment. This was the maximum range of our aircraft and we could not air-refuel as this was deep inside Iraqi territory. Our calculations were by the book and we found out we would have less than 2,500 lb of fuel prior to touchdown at Bushehr. We managed to bomb the trucks and vehicles, creating massive fireballs all around. However, I decided to turn around and fire my plane's nose-mounted gun at other intact vehicles. But this cost me a valuable amount of gas, which caused me a double engine flameout as soon as I touched down on the runway. Had this happened moments earlier, I may have had to eject over the water. This is an unforgettable mission for me.

TOP 11
'JUMP JETS'

Runways are undesirable locations for military aircraft. Being tied to miles of concrete gives jet aircraft a built-in vulnerability as well as restricting their flexibility. So it is hardly surprising that designers have made great efforts in trying to produce vertical-take-off-and-landing (VTOL) aircraft. But creating such an aeroplane is exceptionally hard. These almost inevitably doomed projects have put some fascinating shapes into the sky. Here, eleven of the most notable designs are reviewed in detail by Jim Smith, a former UK special adviser on aerodynamics, weight and performance for the advanced short take-off and vertical landing project that led to the F-35B.

11. VFW VAK 191B (1971)
'The German kestrel'

As with several aircraft on this list, the £192 million VAK-191 was an entry in NATO's huge competition for a supersonic VTOL strike aircraft. The propulsion system, developed with the help of Rolls-Royce, used a Rolls-Royce/MAN Turbo RB.193 (similar to the Pegasus engine in concept) and two lift jets. The aircraft had an internal weapons bay. When the NATO requirement was scrapped (after being technically won by the Hawker P.1154), the VAK-191 flew on for research purposes as part of an ambitious US–West German fighter project. When this project was also canned, it was hard to justify the project, and the VAK-191 was axed by the West German government in 1972.

BEST FEATURES: Simple-to-implement propulsion system, Pegasus approach plus two lift-engines, neat configuration, much higher fineness-ratio than the Harrier.

WORST FEATURES: Low thrust-to-weight ratio on cruise engine and small wing, so wing-borne landing would have been tricky and manoeuvrability poor.

WOULD IT HAVE WORKED?: It made a successful flight, but it would have needed substantial development to become an effective operational system. Requirement became irrelevant, and was eventually filled by the Tornado.

10. Ryan X-13 Vertijet (1955)
'The Pentagon easychair'

OPPOSITE: Sleek West German Advanced Vertifan fighters operate from a dispersed woodland location in this Ryan artist's impression. The propulsion system of this 1965 concept would have been far more complex than that of the only Western Cold War 'jump jet' to have actually entered service: the Harrier.

TOP RIGHT: Theoretically, the simplest way to take off vertically is to have your aircraft sit on its tail, though pilots certainly would not agree. The tailsitter concept may return, just without the pilot, as attested by several unmanned aircraft projects.

One approach to vertical take-off and landing was the 'tail sitter'. The X-13 was more successful than its turboprop tail-sitting brethren but was championing the wrong approach. In an attempt to promote the aircraft, Ryan arranged for the X-13 to cross the Potomac River and land at the Pentagon.

BEST FEATURE: Claimed thrust-to-weight ratio of 1:48 is quite impressive.

WORST FEATURES: Operational concept appears totally flawed. Very difficult to land.

WOULD IT HAVE WORKED?: No. Operationally useless; technology demonstrator only.

9. Lockheed XV-4 Hummingbird (1962)
'Bumbly chancer'

Intended as a target-spotting aircraft for the US Army, this is probably the worst aircraft on this list in terms of its effectiveness. Vertical lift came from thrust being vectored downward through multiple nozzles, but the thrust generated was far less than expected, which is perhaps why the concept moved from vertical to merely short take-off.

BEST FEATURES: First demonstration of an ejector-augmentor VTOL propulsion system, in the XV-4A; subsequent adoption of four lift jets plus propulsion engine, in the XV-4B.

WORST FEATURE: The ejector-augmenter propulsion system produced a thrust-to-weight ratio of only 1:04 – totally inadequate. Both Hummingbird prototypes crashed, one killing the test pilot.

WOULD IT HAVE WORKED?: No. The XV-4A was essentially a failure; the XV-4B propulsion system occupied the whole fuselage, leaving little opportunity for operationally useful load.

8. Yakovlev Yak-38 (1971)
'The Black Sea Harrier'

The much-maligned Yak-38 was only intended as an interim aircraft and shouldn't be judged too harshly. Also, considering it is one of only three VTOL fast jets to have actually entered service, it deserves a little more respect. This equivalent to the Sea Harrier, but with a higher maximum speed, served the Soviet Navy from 1976 to 1991 and laid the foundation for the fast, agile and considerably more impressive Yak-41.

BEST FEATURES: Neatly packaged, two lift-engines plus cruise engine with thrust-vectoring for landing. Automatic ejector seat.

WORST FEATURE: Vulnerable to any lift-engine failure in the hover, hence the need for an auto-ejection seat.

WOULD IT HAVE WORKED?: It was in fact a successful operational aircraft, with about 200 produced.

> ## 'This equivalent to the Sea Harrier, but with a higher maximum speed, served the Soviet Navy from 1976 to 1991'

TOP RIGHT: Yak-38s on the deck of a Kiev-class carrier. Of the four Kiev-class carriers, one was converted into a Chinese theme park then a luxury hotel, one became a Chinese naval museum, one was broken up in South Korea and one converted example (in the Baku subclass) lives on in Indian Navy service as the INS *Vikramaditya* operating MiG-29Ks. The Yak-38 and the Kiev-class carriers were a vital stepping stone to Russian fixed-wing carrier operations.

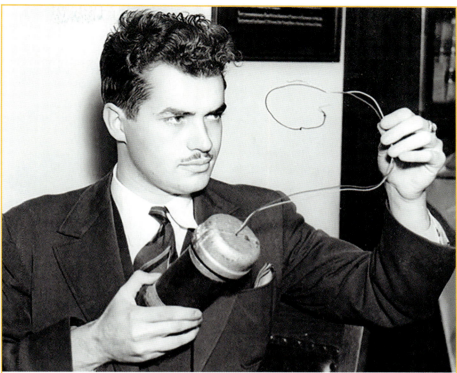

7. Ryan XV-5A (1964)
'The man eater'

The perky little Ryan XV-5A was built to answer the US Army's need for a close-support aircraft. Attempts to develop it into a combat rescue capability were not encouraging, however; in trials, a dummy was ingested by one of the wing fans. The idea of using a lift-fan for vertical flight is still in practice today, and can be seen on the F-35B.

BEST FEATURE: Novel fan-in-wing lift system showing the benefits of moving large volumes of air more slowly. These produced three times the thrust that would have been provided by a conventional nozzle.

WORST FEATURE: The novel fan-in-wing lift system – with its complex doors above and below, and nose-fan for balance and control – proved difficult to manage, particularly in transition.

WOULD IT HAVE WORKED?: It might possibly have worked, but it is doubtful whether the resulting combat rescue system would have been more effective than the rugged Bell UH-1 Huey.

TOP RIGHT: Assisting take-off with rockets was mastered by a team involving 'Jack' Parsons, but it was probably originally a Soviet idea. The most extreme example of rocket assistance was the zero-length launch system or zero-length take-off system.

SEX, MAGICK AND ROCKETS

I height Don Quixote, I live on Peyote,
marijuana, morphine and cocaine.
I never knew sadness but only a madness
that burns at the heart and brain.

Excerpt from an untitled poem by 'Jack' Parsons

For very short take-offs, strap-on rockets can be used, known as jet-assisted take-off (JATO) packs. Research on this use of rockets began in the 1920s, then in 1939 the US government funded an American research effort. A key player in US rocket propulsion and rocket-assisted take-off technology was one 'Jack' Parsons. Born Marvel Whiteside Parsons, he was one of the principal founders of the Jet Propulsion Laboratory (JPL) and was also an occultist and practitioner of 'sex magick' following the outlandish teachings of Aleister Crowley. His accommodation in Pasadena, California, was a hotbed of all kind of strangeness, magick ritual, homebrewed absinthe, Beat poetry, sexual swingers, polyandry and high-explosives development. Among the inhabitants of Parsons' Parsonage was the science-fiction writer and, later, founder of the Church of Scientology, L. Ron Hubbard. Parsons died in an explosion at his home in 1952.

6. Dornier Do 31 (1967)
'Jimbo the ketamine jet'

ABOVE: The superbly bonkers Do 31 transport was conceived to support the off-base dispersal of a planned NATO supersonic fighter that never entered service. It was powered by eight lift jets and two Pegasus Harrier engines. The drag and weight imposed by the wingtip-mounted engine pods was a big issue and the performance was disappointing. The aircraft had a fantastic appearance, however – suitable for Hitler to escape to the moon in.

As with the Royal Air Force, in the early 1960s the Luftwaffe became concerned about the vulnerability of aircraft operating from large air bases. The British developed and eventually deployed the Harrier; the Germans, in a frenzy of innovation, developed and flew – but did not put into service – two potentially supersonic VTOL fast-jets and a VTOL transport, the Do 31. They also experimented with a zero-length launch system for the Starfighter, the ZELL (based on ideas from rocket genius and occultist sex-magician 'Jack' Parsons). As a production aircraft, the Do 31 was envisaged as supplying tactical, logistic support to the fast jets, itself using as forward-operating bases the airstrips on which the ZELL Starfighters were expected to land using arrester gear.

But it turns out that the tactical and logistic support of forward-air operations could be well supported by another aircraft that was in development at the time – the Fiat G.222. This has now been developed into today's C-27 Spartan, which offers similar payload-range performance to the Dornier 31E, albeit with STOL

rather than VTOL capability, and at a fraction of the cost, risk and complexity of a production Do 31.

The Do 31 was an impressive answer to a question that shouldn't have been asked. Technical progress and ambition had run ahead of operational analysis, resulting in flawed requirements.

BEST FEATURES: Integrated flight control and auto-stabilisation system managing two Pegasus engines plus eight lift engines in jet-borne flight.

WORST FEATURE: Weight and complexity of a ten-engine propulsion system.

WOULD IT HAVE WORKED?: There's no reason why this would not have worked, but short take-off utility and cargo aircraft provide a simpler and more rugged solution if the VTOL operational requirement is relaxed.

5. EWR VJ 101 (1963)
'The manga starfighter'

Heinkel and Messerschmitt teamed up with the rather less famous Bölkow to produce this six-engined tribute to the aesthetics of *Roger Ramjet*. Unlike other aircraft featuring small jets, this does not feature a larger main engine, so god knows what would have happened in the event of an engine failure. The design was in many ways similar to the never-completed Bell XF-109.

BEST FEATURES: Six-engine control system in jet-borne flight integrated to throttle (providing collective thrust-modulation), stick and rudder, with roll control by differential modulation of tip-mounted engine thrust and yaw control by differential wing-nacelle tilt. Pitch control by simultaneous differential thrust from two nose-mounted lift engines and the four wingtip-mounted engines.

WORST FEATURE: Use of afterburner on tip engines resulted in ground erosion and hot-gas ingestion problems.

WOULD IT HAVE WORKED?: The propulsion and control system worked. It also achieved a speed of Mach 1.14, which was an impressive achievement. Significant developments were proposed for operational aircraft against changing requirements, which were first met by the F-4, then the Typhoon.

'Unlike other aircraft featuring small jets, this does not feature a larger main engine, so god knows what would have happened in the event of an engine failure'

4. Dassault Mirage III V (1965)
'Saut Mirage'

Without a doubt, the best-looking and fastest jump jet to fly was the French Mirage III V. This prototype fighter, based on the layout of the Mirage III, first flew in 1965 in an attempt to win the NATO Basic Military Requirement 3, for a common supersonic VTOL fighter. Although the prototype aircraft achieved Mach 2.04, it could not fly supersonically after a vertical take-off as it could not carry enough fuel. The aircraft was lifted by a bank of eight lift jets, the volume, weight and complexity of which would have limited the aircraft's practicality had it entered service.

BEST FEATURE: High performance, reaching Mach 2.04 (after a conventional take-off) in September 1966.

WORST FEATURE: Jet-borne flight was achieved through the use of eight lift engines not used in wing-borne flight, resulting in adverse impacts on weight, range and complexity.

WOULD IT HAVE WORKED?: Two Balzac predecessors and two Mirage III Vs were built, but three of these aircraft were lost in fatal accidents. None of the aircraft competing for the NBMR-3 requirement became operational.

3. Lockheed Martin F-35B Lightning II (2008)
'The prolapsing firefly'

Though symbolic of all that is awful about the military-industrial-congressional complex, the F-35, in particular the F-35B STOVL variant, is a very impressive piece of engineering. The F-35B was the first supersonic jump jet to enter service – a highly impressive feat after more than fifty years of failed attempts by some of the world's greatest designers. The aircraft's vertical take-off is fascinating to watch, described somewhat distastefully by one observer as looking like 'a prolapsing firefly'.

BEST FEATURES: Supersonic and STOVL capability with stealth and extensive connectivity. Advanced flight controls and pilot interface.

WORST FEATURES: Long software development and validation delays, and a poor logistic-system performance. Despite high thrust-to-weight ratio, a high wave-drag limits the maximum Mach to 1.6, and the F-35B is also restricted to a 7G manoeuvre limit.

DOES IT WORK?: Significant numbers are now in operational service, but the F-35B has about 30 per cent less internal fuel than the F-35A, due to STOVL mods.

2. Yakovlev Yak-141/41/43/201 (1987)
'Perestroika carpetburn'

The Soviet Union was often accused of stealing US aircraft concepts and technologies, but in reality it was a case of give and take, as well as similar design solutions resulting from parallel teams working to solve similar problems.

That Lockheed bought research from Yakovlev on the STOVL propulsion system of the Yak-141 (a prototype designation that would have become the Yak-41, had it entered service) is pretty notable. The Yak-141, impressive though it was, was merely a stepping stone to the formidable Yak-43 fighter. The Yak-43 would have been far faster and more versatile than the Harrier, with a performance comparable to the MiG-29. In the 1990s, Yakovlev studied a more advanced stealthy version, the Yak-201, analogous to the F-35B.

The tumultuous transitional period that made the collaboration with Lockheed possible also killed the Yak-43, but its DNA lives on today in the F-35B.

BEST FEATURES: Swivelling rear thrust-vectoring nozzle, replicated by Lockheed-Martin for the F-35B. Plus appearance and performance, with a maximum speed of Mach 1.4

WORST FEATURE: Use of afterburning for vertical flight, resulting in noise, ground erosion and potential for hot-gas ingestion.

BELOW: The Yak-43 as it would have appeared in Soviet Naval Aviation colours. This example is armed with R-27 medium-range air-to-air missiles.

WOULD IT HAVE WORKED?: The programme was halted in 1991 due to economic conditions in the Soviet Union, but it would have been an impressive operational capability had development been continued.

'The Yak-141, impressive though it was, was merely a stepping stone to the formidable Yak-43 fighter'

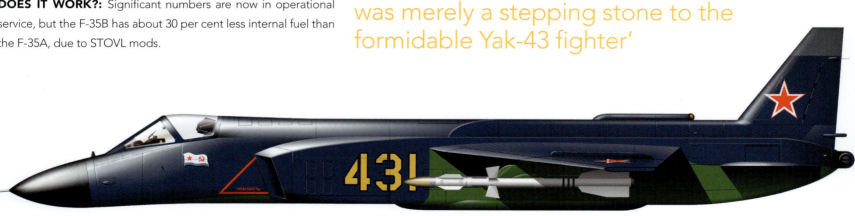

1. Harrier (1967)
'Four-poster deathtrap'

No surprises for guessing the No. 1 spot. The Harrier was the first operational STOVL strike fighter, and the aircraft for which the nickname 'jump jet' was coined. Key to the Harrier's success was the simplicity of the propulsion concept: the engine's thrust is steered through four movable nozzles. Unlike rival concepts, the wing and engine did not need to be swivelled for vertical flight, nor did it depend on extra lift engines (which were a weight burden in forward flight) or a specialised landing pad. The Harrier was a lower-risk brother to the aborted P.1154, initially funded partly by the US Army (which was keen to develop an in-house fixed-wing close-support force) and partly privately, as British companies were then prohibited from developing manned military aircraft (as they were deemed obsolete).

The first-generation Harrier entered service with the RAF on 1 April 1969. Its final operator was the Indian Navy, which operated it, in Sea Harrier guise, until 2016. In the British and American air forces, the Harrier was replaced by the bigger and more sophisticated Anglo-American Harrier II from the 1980s onwards. The Harrier II would also serve with Italy and Spain.

The Harrier was difficult for pilots to master and had a very high attrition rate for an aircraft of its generation (40 per cent of all Harriers were lost in accidents), especially in its initial form. Landing was particularly difficult, with the pilot having to control both the throttle and the nozzle lever with his left hand. Despite these limitations, it was a charismatic and exciting aircraft, sadly missed in Britain, where it retired in 2010.

BEST FEATURES: The simple propulsion concept, coupled with the larger wing and improved structure in the Harrier II, which provided significant range and payload increases compared to earlier variants.

WORST FEATURES: Development was limited by engine thrust and a fat subsonic shape, arguably leading to the early retirement of Harrier and Sea Harrier by the British government. Lateral-directional handling was critical in the hover. Has a high signature as there is no way to screen the engine or reduce the infrared signature from rear aspects.

DOES IT WORK?: All variants have provided sterling service, particularly the Sea Harrier in the Falklands conflict and the Harrier II in a number of operations from the first Gulf War onwards.

VIKING THUNDERCLAP!

INTERVIEW WITH SAAB JA 37 VIGGEN PILOT MIKAEL GREV

Thundering over icy mountains, taking off from motorways and secret caves, the Viggen was tasked with defending a small, neutral country from the biggest war machine in history. It is also a strong candidate for the most charismatic of the Cold War fighters. Former Viggen pilot Mikael Grev gave Hush-Kit his insight into the iconic JA 37, which he flew between 1998 and 2003.

JA stands for 'fighter and strike' in Swedish, but 'fighter' was what it was all about. It is normally just called the Fighter Viggen and it was the best Viggen version – something all pilots agree on. Or else it was the Fighter Viggen pilots that were the best – possibly both! There were also [Mikael flashed a tongue-in-cheek smile here] Recce and Strike versions of the Viggen. There was almost no rivalry among the pilots of the different versions, though.

Acceleration

The Viggen was capable of massive acceleration, at low level, as long as you didn't turn much. The engine had a high-bypass ratio, which means lots of power at low level. On a cold day (we have plenty of those in Sweden), at 30 metres altitude, with 30 per cent fuel left and a clean aircraft, it was like riding a rocket!

Sustained turn-rate

Pretty bad by today's standard. Of course, if you stayed at low level it was OK, but you basically controlled a big air brake with your stick, i. e. the Viggen itself was a large air brake. The first versions actually had a normal air brake, but it was later welded shut since it did nothing compared to the induced drag you could get by pulling a few Gs.

Instantaneous turn-rate

It was OK, but since it didn't have carefree manoeuvring you had to take it a bit easy and keep track of the G-meter. It had a maximum 7-G allowed, and that was with quite low fuel, so it was decent. It had a clever sound-feedback system, with beeping tones for high-G and high alpha (which had a different pitch), so you could pull hard on the stick and still look at the target. But if you wanted that last bit of turn-rate, you had to look at the G-meter with one eye and at the target with the other one – and preferably keep one eye on the alpha meter as well. Marty Feldman would be an awesome Viggen pilot!

High-alpha performance

Being respectful to an old lady, if we interpret this as being able to keep exactly the maximum 23 degrees of alpha in a dogfight, it was fantastic. The flight control system was very good – again, for the time. There was no problem staying between 22 and 23 degrees while rolling and manoeuvring against a target in a dogfight.

If you went over 23 degrees or 7-Gs you would get 'the knife' in your hand (a buzzer built into the stick), which told you to push the stick forward. You would then also get an automatic report after landing, and you didn't want that since it meant 'a talk' with the squadron commander. At 26 degrees alpha, or a bit above that, there was a real risk of super-stall.

There was a trick here that some knew and used. You had to stay above 23 for 2 seconds or more to get a report. Since every degree counts, you could milk it a bit and go back and forth between 24.5 and 23, as long as you didn't stay above 23 for more than 2 seconds. But if you went above 25 you were toast, no matter the duration: you would have to own up to the mistake once you landed.

OPPOSITE: The air defence of Sweden in the late Cold War was entrusted to the JA 37 Viggen. The Viggen had a unique combination of flapped canards and a thrust reverser, and was capable of operating from motorways.

ABOVE: Mikael Grev in the Viggen cockpit. Had war broken out with the Soviet Union in the late 1980s, Viggen pilots would have had the daunting task of facing numerically superior Flankers and Fulcrums. Operating away from vulnerable bases, Biggens would have been able to rapidly rearm and refuel to provide a surprisingly stiff resistance against invaders.

Could the Viggen have survived against a Flanker?

Yes, if we had numerical advantage around 2:1. One-on-one air combat manoeuvring would be pointless, and we'd lose every time. The Flankers take more Gs, have a higher sustained turn-rate and carry more fuel. Luckily, the Cold War never got warm.

Was its lack of agility an issue?

Yes, in a sense. But this was in the mid-nineties and a lot of Swedes thought that eternal peace had arrived (Russia was good now and was never ever going to be bad again), and we had something of a strategic 'time out'. In that way the threat wasn't really real and nothing military was a real issue. Also, the Gripen was already well on its way to entering service and that had much better performance in almost every way.

How did the Viggen concept compare with fighters from other nations?

In Sweden, we do have a very different geographical situation – a large country compared to population and lots of roads. Actually, today we try to get back to the road-base system since it is a really good way to disperse the air force and make it harder to hit with cruise and ballistic missiles. A road base is easy to shut down but it's expensive to open again, so we'll see how many we can get back into use.

Sensor performance and situational awareness

These were good for the time. We had a moving map in the JA 37 C/D, but it was monochrome and vector-based. The Di version, which came in the last years of service, had a colour moving map, just like in the Gripen.

Sensor-wise, it had a pretty good radar, but with a limited range of around 60 km. The Di version had a bit more, due to better processing. The radar warning receiver just told us the quadrant an aircraft locked on from, and we had no forward-looking infrared, missile-approach warning, laser-designator pod or other fancy stuff. I have been told the recce camera was state-of-the-art. Fortunately it wasn't integrated on the Fighter Viggen.

Did you practise off-base operations such as motorway take-offs?

Yes – and the feeling when you stand at the beginning of the runway, ready to take off, with a forest wall a bit over 800 metres in front of you, is not unlike standing on the platform looking down when bungee-jumping. It's the same thing when landing on that same strip – 800 metres seems rather short. It looks like it can't be done, but then you know many have and you just do it.

What was it like using the thrust reverser?

Like falling over forwards while braking on the runway. As a pilot you're used to being pushed backwards, but when it was engaged after touchdown, the forces got reversed. The more you push the throttle forward, all the way to full military power, the more you hung in your straps. It was good to have, especially at the road bases when it was icy, but honestly I wonder if it wasn't more weight to always carry around than it made good the few times it was used.

ABOVE: A JA 37 Viggen armed with RB 24J (the Swedish designation for the AIM-9P3 Sidewinder infrared-guided missile) and RB 71 Skyflash semi-active radar-guided missiles.

> 'The Oerlikon 30-mm was really good, not just for its time. It was actually so good that it was loosened a bit to spread the bullets'

How good was the Skyflash missile?

Not very good compared to the AA-10 Alamo-A or -C that the Flankers had. And with what I know now about missile models and really fast target aircraft, I feel it was very fortunate we didn't have to use them in combat: I currently work as CEO of Avioniq, who create and analyse missile models. In my job I carry out a lot of simulations with different missiles and targets, and I know that the Skyflash missile would have a hard time hitting a very fast target, such as a Su-27 in Mach 2.

What was it like firing the cannon?

Awesome! The Oerlikon 30-mm was really good, not just for its time. It was actually so good that it was loosened a bit to spread the bullets. We had an air-to-air automatic aiming mode for it which used the radar. The pilot just moved the stick in roll and the aircraft calculated and controlled the pitch to hit the target. It was used to fire the gun on larger aircraft with great precision, and if the target was going straight, it hit what it aimed at.

What was the best thing about the Viggen?

The sound was the best. It was a deep, soft sound of pure power, just like a muscle car. I sometimes get asked how the Viggen and Gripen differ, and I usually say it's like a Dodge Charger compared to a Porsche 911.

Describe a memorable flight

I started at 30,000 feet and the aircraft just accelerated and accelerated. It must have been a newly serviced engine in combination with high pressure and cold air or something, because this was more than usual. I throttled back just before feet dry at more than Mach 1.8 and it was still accelerating like crazy! Usually there's no problem reducing the speed with a Viggen, but at that speed there was a feature that kicked in that meant you couldn't go below full military power because of inlet pressure. And you couldn't take many Gs either, because of the high speed. So, land, with houses and stuff, approached much quicker than estimated, since the speed reduction took way longer than expected.

I thought I had broken every window in the Stockholm archipelago. Sitting on the squadron, waiting for the first call to come in, was a memorable part of that mission. But no one called. Not even a peep. I guess the weather conditions were on my side that time.

Three words that describe the Viggen?

Power. Steel. Fighter-Viggen-is-better-than-Recce-and-Strike.

BELOW: The beautiful 'FOA camouflage' is an indigenous vehicle scheme developed in the 1960s for long-range warfare in the Swedish landscape. It is nicknamed *Viggenkamouflage*, as the AJ 37 Viggen was an early adopter. This JA 37 wears a commemorative scheme for its final flight on 9 August 2000, featuring an image of Hagbard, the symbol of 2.FU-komp/F17, then based in Ronneby.

THE GREATEST LIVING FIGHTER ACE

INTERVIEW WITH F-14 TOMCAT PILOT COL. FEREYDOUN A. MAZANDARANI (RTD)

The F-14 was king of the air in the extreme combat of the Iran-Iraq War. Around 180 Iraqi aircraft fell to Grumman's deadly Tomcat, and of these kills sixteen can be attributed to the Iran Air Force's Col. Fereydoun Mazandarani. The world's greatest living ace spoke to Hush-Kit about the perils of full-scale air warfare.

My toughest engagement was with five Iraqi Mirage F1 fighter jets during my annual standardisation and evaluation check while on a special mission flight with Major J. Shokraee-Fard as instructor pilot. It took place near Nowruz oil field, which had been attacked the day before by the Iraqi Air Force. I had actually briefed the pilots that same morning on how the Iraqis would probably attack, i.e. in two groups, one flying at high altitude distracting the CAP fighters while the other group snuck in low to strike the oil rigs.

As had been predicted, we encountered two groups heading our way from two directions: a flight of two and a flight of three. As soon as we prepared to engage the enemy, at 690 knots and slightly over 50 feet above the water, I noticed that our Master Arm switch had failed, leaving us defenceless. The hunter had become the hunted. The attacking Mirages fired six air-to-air Matra missiles at us. Making hard turns and pulling high Gs, we defeated the missiles and re-engaged them in a canopy-to-canopy dogfight. We were so close that in a couple of passes I could see the pilots' white notepads strapped to their legs.

Major Shokraee-Fard kept checking our six, advising me of enemy position while I kept manoeuvring hard, keeping myself out of their gun or infrared missiles lock. During one of these manoeuvres we saw one Mirage crash into the water while the others returned to base. Once we were clear, I noticed that my G-suit had ruptured from the pressure and my helmet had cracked hitting the canopy. On our way back to base, we were advised by ELINT and the local ground radar that only three of the five Mirages had returned. After the flight, Major Shokraee-Fard had to wear a neck brace for six months, while I suffered injuries to my knees that resulted in two operations after my retirement. The G-meter was locked at 11.5 Gs on the gauge, which required the Tomcat to go through non-destructive inspection. The analysis showed nineteen cracks and fractures along the longitudinal axis of the aircraft, which put it out of service for almost two years. We were really lucky that day.

In the early days and weeks, the high losses of our pilots in the F-4 and F-5 squadrons were especially hard and painful, affecting the overall morale. It was quite bleak. As the days went by, we realised that the only available force that could slow down the rapid advance of the Iraqi ground forces was the air force. After a few weeks, despite the repeated loss of our colleagues, the missions continued without any problems and the bitter realities of war became routine. We had no choice. Iran's ground forces were in disarray after the revolution, as a result of widespread purges – and in many cases they were no match for the Iraqi onslaught. Therefore the air force took it upon itself to act as speed bump against Iraqi ground units until our own soldiers could be organised into an effective fighting force. We performed close air support while providing barrier combat air patrol [a mission flown between an area or force and the direction an enemy attack is most likely to come from] to our own cities and infrastructure.

My biggest high was to be the first Iranian Air Force pilot to have done a night refuelling in an F-14. We were not trained to

OPPOSITE: Col. Fereydoun Manzandarani beneath the aircraft that he took to war, the formidable F-14 Tomcat. Like every Grumman carrier aircraft, the F-14 was built like a tank.

LEFT: The Grumman-developed Tomcat logo was this cocky gunslinging cartoon character. Like the F-14 itself, the cat has twin tails and a gun on the left side.

ABOVE: Iran's F-14s were arguably the most successful of all late-twentieth-century fighters. Only the F-4 in US use or the F-15 in Israeli service can approach a comparable combat record.

do this by our former US Navy instructors so I was quite proud of myself for doing something like that. The biggest low would be losing three F-14s within a short few days to the French-built Mirage F1 used by the Iraqi Air Force. That hurt our pride badly.

I had eight aerial kills with the Phoenix missile, two kills with the Vulcan M61 A gun and one kill with the MIM-23 Hawk missile that we ended up using on our fleet of F-14 jets due to severe missile shortages in late stages of the conflict. On top of that I can claim five manoeuvre kills from two separate engagements.

My first air-to-air kill was a few days before the official start of the war, on 17 September 1980, when two Iraqi MiG-21s were on a bombing run over the city of Mehran in western Iran. The ground radar guided us toward the Iraqi fighters and we approached them in such a way that I ended up 13–15 miles behind one of them at 3,000 feet. I was flying with my RIO Lt Sultani, at over 240 knots faster than the Iraqi fighter, when I fired my Phoenix missile and watched an aircraft disintegrate and explode visually for the first time.

My second and third kills were with RIO Lt Najafi on 25 September 1980, flying over the general area of Dezful in south-west Iran. We were informed by ground control radar that four MiG-23s had crossed the border. We were directed towards them and at around 45 nautical miles [NM] locked our radar on the fighters. They reduced their altitude and turned towards Iraq, flying in between the mountains. At one point my RIO noticed that he only had two radar signatures and, when we checked with ground radar, they also confirmed two targets. My sixth sense told me that this was a ruse and that only two of the fighters had actually returned. I turned the plane around, and within mere seconds reacquired the missing two hostile aircraft flying at 100

feet and heading for the city of Yasouj. We immediately locked on both targets and launched two Phoenix missiles. One of the MiG pilots survived ejection and was later captured.

My fourth kill was a MiG-23, on 13 November 1980, my backseater being Lt Ahmadi. We were providing top cover for CAS missions, and the MiG was in pursuit of two Iranian Air Force F-5E Tiger jets returning from a CAS. Our position didn't give us enough time to engage him with the Phoenix missile so we prepared ourselves for a knife fight. We both began turning into one another, trying to get each other in our respective gunsight. We began spiralling downward in a rolling scissor manoeuvre. I opened fire with the gun twice, but didn't think he was hit. I told my RIO to keep reading the altitude as we hurtled towards the earth. I kept hearing him read the altimeter: '2,500 feet, 2,000, 1,800, 1,500, 1,000, 600, 300' – and then I pulled the nose up hard, pushing the throttles to Zone 5 afterburner, avoiding the ground. The moment I levelled off, I inverted the plane in time to notice a fireball on my left side. The MiG had impacted the terrain. Honestly, I really would have wanted to meet with this skilful pilot and I felt bad about his sudden demise.

My fifth and sixth kills were on 29 November 1980, a day after the Morvarid naval operations that decimated the bulk of Iraqi naval forces in the northern Persian Gulf. My RIO on this day was Lt Ibrahim Ansareen. The main objective of the operation was to destroy Iraq's Al-Bakr and Al-Amaya oil terminals in order to cut off Iraq's oil exports. The enemy naval surface forces had lost several fast attack boats, on top of a few Russian-built corvettes, to our well-equipped navy, so the Iraqi Air Force was tasked with providing cover for their helicopters trying to retrieve Iraqi troops and sailors lost at sea or still on the oil platforms. Our job was to deny their top cover and also target their helicopters as they came in to reinforce their positions.

I took out one Iraqi MiG fighter aircraft with a Phoenix missile and continued playing cat-and-mouse with the remaining fighters. At an opportune moment I launched a second AIM-54A Phoenix missile at another approaching MiG, about 10 or 11 miles out, resulting in a shoot-down. Since my aircraft was loaded with only two rounds of AIM-54 Phoenix missiles, and therefore we were practically 'Winchester' [out of weapons], I immediately called for support. Two of my buddies, Capt. Jamshid Afshar and the late Capt. H. Farrokhi, who was patrolling in another area, came to replace us. Those two men shot down three other intruding Iraqi fighter jets, resulting in total shoot-down of five enemy aircraft by our F-14 fighters in a single day.

My seventh kill happened on 24 April 1981, when Lt Farrokh-

Nazar and I were providing cover for our ground troops near Ahvaz. We were vectored by ground radar to intercept an aircraft heading for our friendly positions. We tracked him on our radar, and at about 20 NM fired a Phoenix missile, resulting in a shoot-down.

My eighth and ninth kills happened around 20 November 1982, with Lt Tahmasebi as my backseater. We received a NOTAM ['notices to airmen' contain urgent, last-minute information] about the arrival of Saddam Hussein and/or a few high-ranking officers to inspect the Iraqi battle areas. On that fateful day we could see higher-than-normal Iraqi air activities, which supported our initial intelligence about the presence of high-ranking figures in the battle area. As usual, we'd lock on to the Iraqi fighters a few times and they would immediately turn around; however, during one of these radar locks the Iraqi fighter and its wingman continued its course towards us. Knowing who they were and what they were going to do, we fired two Phoenix missiles at them but could not really tell if those launches were successful. The moment we landed, we were summoned by our angry wing commander. We were both given a written reprimand for not following the rules of engagement [ROE] as we had been directed. Apparently, our ROE had directed us to engage the Iraqi fighters only inside Iranian territory. We were also forbidden from crossing into Iraqi air space by our headquarters. This was set up to prevent the loss of a prized F-14 aircraft with

'We were patrolling in Sector 3, close to Minoo Island, south of the city of Abadan, when we encountered a single Mirage flying low'

its sophisticated weapons system inside the enemy territory in case of a mishap or loss to enemy fire. The air force brass had also feared our top-secret AIM-54A missiles would somehow end up in enemy hands if they were fired at enemy fighters inside the enemy air space. As a result, we were not credited for these two kills. However, thirty-one years later, in 2013, IRIAF's centre for study and research officially registered these two kills in our names.

My tenth kill was in the north-east of Boubyan Island, with Lt S. Shokouh, in February 1984. The target was a Mirage F1 that we destroyed using the nose gun. We were patrolling in Sector 3, close to Minoo Island, south of the city of Abadan, when we encountered a single Mirage flying low. We didn't have enough time to prepare a radar lock. I managed to get myself to his four o'clock. Switching to PLM (pilot lock-on mode) on the radar, for close combat or dogfight, I locked on to the Mirage but the pilot immediately went into full afterburner and began jinking,

ABOVE: The AIM-54 Phoenix was long the most capable air-to-air missile in the world. Though capable of use at extreme ranges, it was too precious to risk being used at the edge of its envelope. Unleashing it at shorter but still sizable distances insured an extremely high probablility of kill.

ABOVE: Iranian F-14s over the mountains of Iran. Early F-14s such as these featured glove vanes that retracted at subsonic speeds and extended automatically when going supersonic.

'Another myth that needs to be shot down right now is in the myriad statements by F-14 enthusiasts boasting about shooting down Iraqi fighters from a distance of over 150 km! That is untrue'

preventing us from getting on his six (the six o'clock position, meaning directly behind).

I managed to manoeuvre myself to about 25 degrees from his tail using afterburner as we flew below 100 feet AGL. At a distance of 500–700 feet I opened up at him. It took two bursts of the F-14's Vulcan M61 A1 cannon to effectively strike the Iraqi fighter. The Mirage caught fire and unfortunately the pilot didn't have enough time to eject at that high speed and at such a low altitude. The burning fighter slammed into the ground and exploded.

My eleventh kill took place on 16 September 1986 during the live firing test of the US-made MIM-23 Hawk surface-to-air missile carried by the Iranian F-14. Some quick background should be provided here. The late Shah's air force had ordered plenty of AIM-7F Sparrow and AIM-9L Sidewinder air-to-air missiles, along with

the AIM-54A missiles for our new fleet of F-14s. But the hostage crisis of 1980 effectively killed the prospect of the arrival of the former missile types, thus the Iranian F-14s and crew went to war with only the AIM-54A being the F-14's standard missile. The two versions then available in our inventory, namely AIM-7E and AIM-9J, were not really compatible with the F-14's fire-control system. The MIM-23B Hawk SAM was the only medium-range anti-aircraft missile in our inventory. We adapted it for our F-14 and called it AIM-23C Sedjil. I was a senior member of the team that worked on converting the Hawk surface-to-air missile to an air-to-air missile. During the final trial run, I was ordered to test the missile during an actual combat situation to prove the system to naysayers. I was sent on temporary duty travel to Bushehr 6th tactical air base to stand at scramble alert. I think it was on the third day when the opportunity arrived to demonstrate the combat capability of this new weapon. My backseat was 1Lt Ansareen.

We locked on and fired the first Hawk, which turned out to be dud. The missile made a barrel roll over the nose cone of the F-14 and fell straight down. I immediately fired a second one at 20 NM, resulting in a positive hit as confirmed by our SIGINT [signals intelligence] and radar data. I am told the target was a French-built Dassault Super Étendard maritime strike aircraft leased to the Iraqis in the mid-1980s. These maritime strike jets were used extensively by the Iraqi Air Force to strike our cargo ships and oil tankers, using their infamous Exocet missile. And, as you know, the Étendards had a rather stellar record during the Falklands War against the Royal Navy surface assets, as well.

What is the biggest myth about the Iranian F-14s?

The most tiresome is that the departing US personnel stationed in Iran managed to sabotage Iranian F-14 radar, electronics and Phoenix missiles before leaving Iran in the ensuing days after the 1979 revolution. Let me tell you that I was a young officer during those days at Esfahan Khatami air base, and our wing commanders and senior officers made sure this never happened. We lined up departing American personnel before they boarded their TWA aircraft and inspected them all. Another myth that needs to be shot down right now is in the myriad statements by F-14 enthusiasts boasting about shooting down Iraqi fighters from a distance of over 150 km! That is untrue.

What was your most memorable mission and why?

One of my most important and stressful engagements was on Thursday 21 March 1985, when I was partnered with RIO Lt Sanatkar. I was on final approach, returning to base from a 6-hour combat air patrol. Wing command ordered us to cancel approach and fly back south as fast as we could due to a lack of available covering fighters in that critical sector. I air-refuelled and headed south towards the Khark Island area to offer aerial protection for a con voy of some twenty massive oil tankers fully loaded with Iran's month's-worth of crude oil output. We were on station for more than an hour and nearing bingo fuel status, with only 8,500 lb left in the tanks. I was about to head to our tanker for another mid-air refuelling when the radar controller warned us of ten approaching bogeys.

It was not long before these bogeys became bandits. But the problem was that my aircraft was quite low on fuel for any engagement. We turned our F-14, initially heading south, but again turned to heading 300 degrees as thirteen targets began appearing on the radar scope flying at 500 feet above sea level. (We later found out they were eight upgraded MiG-23 bombers with five Mirage F1s acting as escort.) I had no choice but to engage the enemy fighters in order to protect the convoy of tankers from a horrendous air assault. I told my backseater to be ready to bail out whenever I instructed him to do so. I set my altimeter to 35 feet above sea level and thought out my engagement strategy as I began dropping altitude.

At about 20 nautical miles, with our fuel at 2,000 lb, the escort fighters fired some twenty Matra missiles at us unsuccessfully as the onboard ECM, chaff and flares began to work their magic. We

started defeating the missiles by jinking hard and turned into the Iraqi strike package. Fortunately, the MiGs jettisoned their payload and broke formation as we manoeuvred between the enemy fighters and frightened them. The escort fighters, which were flying in two groups, had also broken formation, and as I passed the last three Mirages I banked hard and got behind them, watching the Mirages light up their afterburners as they headed back. I thought of chasing them but my fuel gauge was now reading 600 lb almost 50 NM south-west of Khark Island, so I disengaged and eased back on the throttle to military power, returning towards Bushehr air base.

I asked Lt Sanatkar if we had been hit or anything, but he confirmed that all systems were working just fine. I called the ground radar but didn't receive a reply. I tried again and still got no reply. After my third call, the operator replied, asking if we were still in our aircraft! They had assumed we had been targeted by all those Matra missiles that had run their course and exploded mid-air. I told them that we might have to ditch the plane and might require immediate search and rescue helicopter overhead. But right then I was interrupted by my dear friend Col M. Reza Moharrami (nicknamed Mamish), the pilot of a KC-707 fuel tanker, saying that we would soon be shaking hands. To our astonishment, this brave marvel of a pilot had flown towards the engagement area, faking radio communications with other aircraft to give the enemy side the false belief that more friendly fighter aircraft were headed to my engagement zone. We hooked up with the tanker at 2,000 feet and gradually climbed to 22,000 feet, receiving much-needed fuel in the process. On our way back to base, I was informed by ground-based radar that one 'Flogger' and two Mirage F1s had not been able to make it back, which was the icing on the cake!

ABOVE: Much to the US's chagrin, Iran still uses F-14s today. It is a cautionary tale of the danger of exporting your best weapon systems to foreign operations.

CUTTING IT IN THE 'ELECTRIC CAKE SLICE'

INTERVIEW WITH A MIRAGE 2000 PILOT

After mastering the Lightning and the Tornado, the RAF's Ian Black volunteered to fly France's hottest fighter, the superb Mirage 2000, which he did between 1993 and 1997. Even after flying the mighty Lightning, the 2000 remains Ian's favourite aeroplane. He explains what it was like to fly the ultimate Mirage.

How did you end up flying Mirage 2000s?

I'd flown Air Defence for around twelve years and converted from back seat to front. I'd reached a point in my career where I had to expand my horizons. I could go down the staff officer route, apply to the Red Arrows, to Test Pilot School or to try for an exchange posting. I opted for the exchange option as I wanted to fly an aircraft the RAF didn't have, and the opportunity to learn a foreign language appealed, too. At the time, the RAF had exchange postings for Air Defence pilots on the F-15C/F-16/F-18, F-4F and Mirage 2000. I wanted the French exchange because it was based in Provence and the Mirage is a unique airframe.

Which variant did you fly?

I flew the Mirage 2000C – the RDI variant – at the time the French Air Force had the Mirage 2000C RDM [a pulse radar] and the RDI pulse doppler radar. They also operated the Mirage 2000D and 2000N. Eventually a Tornado GR1 pilot flew the Mirage 2000D,

but the 2000N's nuclear role meant no foreign pilots were allowed to operate it.

What were your first impressions of the cockpit?

Slightly disappointing at first – I'd come from the Tornado F3, which was painted grey, then blacked out for NVG work, and was very spacious and well laid-out. The Mirage 2000 is more like a fighter from the seventies, with a lot of analogue displays. The rear view wasn't as good as an F-16 and it was pretty cramped. On the plus side, it wasn't overly complex.

Is it easy to fly?

Yes and no. It's easy to fly once you get the hang of it, but the delta wing takes a unique approach to flying – it's not like a conventional wing. It generates huge amounts of lift but also an enormous amount of drag – great for a 'bat turn' but you always end low on energy afterwards. Landing is pretty straightforward; the view is good; air-to-air refuelling is easy. It has very well-

ABOVE: Ian Black leads a fourship of Mirage 2000Cs. According to Black, the nearest aircraft is piloted by 'Nayot', a pilot in the French equivalent of Top Gun (*Les Chevaliers du ciel*); the next aircraft is Siegfried Usal's (he would become right-hand man to President Sarkozy); and the fourth is flown by solo 2000 display pilot Rallet.

balanced controls and gives you great seat-of-the-pants-type senses, with great feedback to the pilot using its early fly-by-wire controls, without feeling like a computer game. I'd almost say it was the perfect blend of old and new.

What is the hardest thing about flying the Mirage 2000? Any quirks?

As I've said, the delta wing could catch you out – it would give you more than 9-G performance but at a penalty; flying in the circuit could be a challenge, and turning finals required quite a lot of pulling on the stick, which loaded the wing up as the drag built. Once you rolled wings-level, it was imperative to take the power off or you would accelerate too quickly.

'Dassault make fine aircraft; and, apart from the ejector seat, it pretty much is 100 per cent future-proofed'

How does the acceleration and climb compare to a Lightning?

The Lightning had two massive Rolls-Royce Avon engines; the Mirage 2000 had one, but it was still pretty potent.

What was the most challenging fighter you faced while flying the Mirage?

Probably the F-15C, as AMRAAM was just coming into service, which totally outclassed us. They had amazing SA [situational awareness] and the way they operated was impressive.

How would you rate the M2000 in the following:

Instantaneous turn-rate: Stunning at all altitudes. With its big wing, even at 50,000 feet, using the leading-edge slats, it could still turn well.
Sustained turn-rate: Sustained turn was still good, especially at low level, where you had sufficient energy to maintain speed.
High-alpha performance: The Mirage 2000 was legendary at its low-speed high-alpha passes – 120 knots was pretty easy to fly.
Weapon system: As a weapons system, the Mirage 2000 is a great 'package', with a good radar, onboard electronic countermeasures and radar warning receiver. It also packs a good array of weapons – with air-to-air refuelling, it's a formidable fighter.

Which three words best describe the M2000?

Vive La France! It's *sexy*; it's *French* – Dassault make fine aircraft; and, apart from the ejector seat, it pretty much is 100 per cent *future-proofed* – the M2000 first flew in 1978 and it's still in service in 2022; despite its sleek frame, it's built like a tank and can pull 9-G all day long.

How would you compare the aircraft to an F-16?

I'd say the F-16 has the edge: while the M2000 evolved from the RDM to the RDI and RDY [a multi-target doppler radar]

versions, they were pretty small upgrades in terms of airframe performance. The latest-block F-16s are a world apart from the original F-16As. Part of the Mirage 2000's problem was the arrival of Rafale, which pretty much stopped any further Mirage development.

How does it compare to the Tornado F3?

The Mirage 2000 is a fourth-generation fighter and extremely capable in both air-to-air and air-to-ground roles, as well as being highly manoeuvrable even when loaded up. The Tornado was extremely competent at the role of interceptor but lacked the agility of Dassault's masterpiece.

What was your most notable flight on the Mirage 2000?

When you fly a Mach 2.0-plus 9-G fighter, trust me: they are all pretty notable. A few stick out: night missions with air-to-air refuelling over Bosnia or live missions protecting high-value assets over Iraq were pretty noteworthy. Flying in another air force's aircraft is a real honour – the trust they have in you is humbling.

ABOVE: The Mirage 2000's large wing enables a dizzying instantaneous turn-rate. Like other French aircraft, the Mirage 2000 features a non-retractable 'walking stick' air-refuelling probe.

INDIA'S HEAVYWEIGHT HAMMER

INTERVIEW WITH FLANKER PILOT ANURAG SHARMA

Described as a 'royal, merciless game-changer', the Sukhoi Su-30 Flanker is a monster: a long-range, well-armed, unbeatably manoeuvrable fighter uniquely equipped with 3D thrust-vectoring control (TVC), enabling it to perform seemingly impossible aerobatics in the sky. We asked Indian Air Force Group Captain Anurag Sharma to tell Hush-Kit more about flying and fighting in the Russian superfighter.

'The Su-30MKI has perhaps as many players as the Typhoon. The Russians provide most of the hardware; Indian, French, Israeli industries provide software, avionics and weapons'

Relative manoeuvring, in relation to an adversary in the sky, requires extensive training and skill development. The manoeuvres can be counterproductive if not done correctly.

noise and sheer power of the accelerating missile is breathtaking.

How confident would an Su-30 pilot feel going against a modern US Air Force F-15C?
As far as the platform is concerned, he's got a better baby in his hands, no doubt!

What is the greatest myth about the Su-30?
That it's too big to manoeuvre.

How combat-effective is the Su-30?
It's a game-changer!

How reliable is the Su-30 and how easy is it to maintain?
Reliable – yes! Maintenance – extensive!

How much post-stall manoeuvring can the average pilot do? Is this a rare skill?
Independent manoeuvres? They do it from day one – it's that easy!

What is the hardest manoeuvre to pull off in an Su-30?
A downward combat manoeuvre with thrust-vector control at low levels, against a manoeuvring target.

What tips would you give new pilots coming onto the Su-30?
It's like a *tapasya* (a Sanskrit word meaning total, selfless commitment). So, dedication, commitment and patient hard work will reveal to you the true pleasures of flying! Early days are tough; just hang in there, get over the hump and you will experience the heavenly pleasure that only fighter pilots have been blessed with.

Tell us something we don't know about the Su-30
The Su-30MKI has perhaps as many players as the Typhoon. The Russians provide most of the hardware; Indian, French, Israeli industries provide software, avionics and weapons. The Russians won't give their knowhow to Israelis and the French won't give it to Russians. So it's is a great achievement to get these components talking to each other. The heart of the avionics system that communicates with all these various systems is Indian.

ABOVE: Indian Su-30s 'fought' RAF Typhoons in several exercises. Clashes of such high-performance fighters garner much attention and speculation; both fighters have different strengths and weaknesses, with the Su-30 boasting extreme supermanoeuvrability, which gives it an edge in low-speed 'dogfights'.

BELOW: This Indian Air Force Su-30MKI is armed with R-27R and R-77 missiles.

LA BÊTE ELECTRIQUE!

INTERVIEW WITH DASSAULT RAFALE PILOT PIERRE-HENRI 'ATÉ' CHUET

Taking off from the perilous deck of an aircraft carrier, Pierre-Henri 'Até' Chuet flew the Dassault Rafale M into combat in Iraq. Hush-Kit asked him to tell all about the Rafale – a remarkable fighting machine, a masterpiece of design and a strong contender for the title of best combat-aircraft all-rounder.

What were your first impressions of the Rafale?

It's a space shuttle! was my first impression. It is very agile, very responsive when you're lightly loaded and very, very manoeuvrable – you can easily bump your head; I bumped my head twice on the first flight! Flight controls are very different as you can barely move the stick; it travels by just centimetres compared to the former flight-control system of the Super Étendard, so it took me couple of hours to get used to that. That's the big difference. A lot of fun on that. . . Another first impression was the thrust, speed, comfort – the fact the aircraft was really sanitised for sound so you have no clue what speed you're flying at; you really have to look at the instruments. And it is extremely responsive.

How would you rate the Rafale M in the following categories?

Instantaneous turn-rate, sustained turn-rate and high-alpha performance: It's good – it's very good. You have two ways of flying the aircraft: you have the air-to air mode, where you pull 9-plus Gs up to 11; then you have it with bombs and full tanks, when your performance is not as good – about 5-plus Gs and about 200 degree less roll-rate. So, it's two different aircraft. When you're in air-to-air, instantaneous turn and sustained turn are pretty good. High alpha is less than a Hornet, but still good – it depends what you make of it. I've never had any issues.

Acceleration and climb-rate: The acceleration is insane! Climb-rate is firm; to give you an idea, if we're at 500 knots and 500 feet, put the afterburner on, wait for the afterburner to kick in, then put the nose up at 60 degrees so you feel like you're vertical because of the angle of the seat (that's 30 degrees), and at some point you have to throttle back in the afterburner to make sure it doesn't go supersonic – in the climb, 60-degree nose up!

As a carrier aircraft? And as a carrier aircraft it's a good jet. Very versatile, very robust, you have enough fuel and it's pretty fuel-efficient. You're burning less fuel in afterburner at high altitudes than Typhoon does without the afterburner.

Against a Super Hornet? 'Honestly, the issue is comparing aircraft all the time. Life isn't that easy. Combat is unfair. It's never going to be fair. It isn't designed to be fair. If you get into fair close combat, you're a bad pilot. Don't put yourself in a fair fight in real life, as that's stupid. Manoeuvre – take advantage and surprise your enemy. It's not about one individual defeating an enemy; you're here to get results. We are result-driven personnel. It's not all about me. You've got thousands of people building a

TOP: Chuet in the cockpit of the remarkable Rafale. Over his shoulder is an F-16.

ABOVE: A Rafale two-ship. Apparent in this view is how close the canard foreplanes are to the main wing compared to other types like the Typhoon.

'Look at the Messerschmitt 262 back in the Second World War: most of them got shot down on landing. An aircraft shot down still makes the count'

ABOVE and RIGHT: The Rafale's configuration, with coupled canard and wing, is optimised for maximum lift generation and ordinance-carrying capacity across a wide range of speeds and angles of attack.

Rafale, and building and maintaining a carrier. There's thousands of people making sure I can take off – if I want to go fair-against-fair, I'm stupid. What I want to do is make sure I win. Why do I say that? If I'm going to fight against a Super Hornet, I'm going to find a tricky way to defeat him. Look at the Messerschmitt 262 back in the Second World War: most of them got shot down on landing. An aircraft shot down still makes the count. If we have to face the US Navy, it's going to be disproportionate in terms of numbers – it's going to bring tactics to an entirely new level. Now, if you want me to do a fair 1-v-1 fight with a Hornet in close combat, actually I'd rather face a Super Hornet; I find the C to be more manoeuvrable than the Super Hornet. As a Rafale, we have an advantage over a Hornet as well. What I would be wary of is their AIM-9X and their helmet visors. So I would be very careful about that.

The Rafale and Typhoon are often compared. How confident would you be fighting against a Typhoon? And why?

I don't know why they're compared so often – it's really not the same design, ideas or philosophy. We're a truly omni-role platform. Typhoons are great; they like to use their big engines at 40,000 feet. I can't count how many times I've shot down Typhoons at 45,000 feet in the contrails – and with my radar off, everything off, from 100 feet below, supersonic in the climb from below, absolutely undetected. So I have absolutely no fear of the Typhoons. Both the tactics used by the Typhoons, the agility and the cockpit of the aircraft make it easier for us to have the advantage – basically we have better fusion of the sensors, so we can be way more aggressive in terms of tactics. It's a great aircraft at high level, but we're not dumb enough to try to fight Typhoons at 45,000 or 50,000 feet. We're going to put them outside their comfort zone; against devious tactics. Now, if you want to rate a Typhoon with AMRAAMs against a Rafale at 50,000 feet, then, yeah, the Typhoon is going to have better performances for sure. But as a Rafale pilot, I'm stupid if I take him on like that, so I'm going to move the combat a bit. I'll fake a combat at 50,000 feet and I'm going to send a guy sneakily low-level to surprise the Typhoon. It's easier than you think!

Which aircraft have you flown DACT against?

Against F-16, Typhoon, Super Hornets, Harrier, Alpha Jet and

ABOVE: Though the Rafale has been designed for reduced radar cross section and features serrated panel edges typical of stealth aircraft, its external weapons and fuel carriage are far from stealthy.

Mirage 2000. The Harrier can really turn around pretty fast, so you have to play it very close; you have to be careful with that. And with the Alpha Jet, don't go into a slow fight with it. It can manoeuvre and do some barrel rolls at pretty low speeds so you really want to pay attention. You can easily be tricked at low speed by an Alpha Jet, so you want to keep your energy high.

The Rafale's cockpit seems very snug. Are there large Rafale pilots?

We do have larger Rafale pilots! And trust me, when you come from the Super Étendard, you find the cockpit to be large. So really, no concern about that.

Something we don't know about Rafale?

The environmental control system is loud as fuck! You lose the ECS and it gets so quiet you think you have a two-engine fire! It happened to me once.

Do you feel confident flying against modern air defences in a non-stealthy aircraft?

Great question. I'm not sure an aircraft's stealthiness is going to make much difference anyway against very modern stuff. We're not afraid of low-level penetrations in the French Navy and Air Force. So, come and get me with your S-400 if I'm at 200 feet above the ground – that's not going to happen anytime soon. And if you want a lot of munitions or stores, you're going to lose on your stealthy signature anyway. So it's not something of much concern; that's why we train to keep current at very low-level penetration. Which is really good as we get to fly at low level, which is awesome. I can't complain!

How confident would you be fighting a F-22 Raptor in within-visual-range (WVR) air combat?

Obviously you have seen videos [of the Rafale doing well against the F-22 in close-in dogfight training]. Is it going to be guns only? Is it going to be Sidewinders? If it's guns only, I don't have any issue; if it's Sidewinders – and he has his helmet-mounted stuff* and 9X – then I'm going to be careful; I would be concerned. I definitely don't have any concerns otherwise, though it would be tougher for me because he has his 9X and mounted visor. If I play my cards correctly, there's no reason why it shouldn't be OK. I have questions, like what is the set-up? Is it going to be

---

* Editor's note: as far as I know, operational Raptors had not been fitted with HMS at the time of this interview.

> 'I like the design of Rafale aircraft a lot, though I must confess I find the Mirage 2000 very good-looking as well, and slimmer and maybe faster-looking'

'butterfly', with one close to the other one? It really depends on the situation. But guns only? Honestly, no concern. And it's a big aircraft, so it's easy to shoot at.

The Rafale is described by many as the most beautiful fighter in production. How do you rate its aesthetics?

I like the design of Rafale aircraft a lot, though I must confess I find the Mirage 2000 very good-looking as well, and slimmer and maybe faster-looking — and it is faster than the Rafale: we're talking Mach 1.8 against 2.2. But I think the Rafale is a good-looking aircraft – then again, it's like asking a dad if he thinks his kids are good-looking or not! So we're biased anyway. But compared to the Typhoon, you can tell it's a good-looking aircraft. I like the Hornet's shape – I think that's a good-looking aircraft too – and the F-22 is one of my favourite-looking aircraft! The F-35? I really don't like the design – I think it's a shitty-looking aircraft, to be honest – but don't quote me on that!

OPPOSITE, TOP: The Rafale is one of the most effective multirole fighter-bombers in the world. Large external fuel tanks increase its already impressive reach.

OPPOSITE, BOTTOM: Deck life is hard on aircraft, and carrier aircraft need tough, beefy undercarriages. The naval M has a far stronger undercarriage than the air force's C variant.

ABOVE: The Rafale is the latest in a long line of attractive French aircraft. This Rafale M wears a commemorative scheme for the seventieth anniversary of Flottille 12 F.

MODERN FIGHTERS

As terrible recent conflicts have shown, the fighter aircraft is far from dead, and can prove vital for a nation's survival. The most modern fighters can shoot down an enemy aircraft at a range of 100 miles, perform precision attacks against multiple ground targets simultaneously and grant their pilots a godlike level of situational awareness across thousands of square miles of ground, sea or air. Modern fighters need to be extremely potent, as they are massively expensive, and so available in ever smaller numbers. Where a fighter airframe of the Second World War could expect a service life of a couple of years, today's combat aircraft designs often soldier on into their forties. Remarkably, the majority of modern fighter types in service today have their roots firmly in the Cold War. Some fighter types emphasise stealth, others performance or durability, but each is ingeniously conceived for the same deadly purpose: destroying an enemy as quickly as possible.

EUROFIGHTER TYPHOON

The Typhoon is currently the best-armed Western fighter for beyond-visual-range combat, bar none. It is very fast, high-flying and energetic, imbuing its AMRAAMs and Meteor missile with a longer reach than those launched by lower-performance aircraft. Its greatest weakness remains its lack of an AESA radar (though Kuwaiti Typhoons do have AESA) and its non-stealthiness. However, the mechanically scanned radar carried by the majority of Typhoons is a decent size, with a good detection range, and is fully mature. The Typhoon's PIRATE infrared search and track sensor is probably the best of its kind, and is capable of emission-free detection and tracking of targets at useful ranges. Though excellent for its day, the cockpit is a generation behind the large-screen F-35 and Gripen E.

Wild turn-rates, a true 9G performance and enormous excess power make the Typhoon a hell of a dogfighter; the highly regarded G-suits worn by Typhoon pilots increase tolerance to the high G-forces. The combination of advanced missile and advanced helmet system give the Typhoon a terrifying off-boresight shot-capability. The Typhoon's main performance limitation is a pedestrian high-alpha performance, but the Typhoon is not an 'angles fighter' like the F/A-18, which relies on risky but startling attacks in the merge; the Typhoon is an 'energy fighter', using its phenomenal ability to preserve energy in a dogfight to wear its opponents out. In short, if an opponent doesn't get a Typhoon on his first attack, he is in a very dangerous position.

F-22 pilots who have 'fought' the Typhoon in DACT have been impressed by its energy levels (especially in the first turn) and several Luftwaffe Eurofighters returned from exercises proudly displaying Raptor 'kill' silhouettes beneath their cockpits. Like the Raptor, the Typhoon has such a formidable reputation that any 'victories' against it in training exercises make excellent boasts. In close-in training fights with a Raptor, the Typhoon has a brief window of time in which it can use its helmet cueing system and superior infrared missiles to gain a 'kill'; once this moment passes the thrust-vectoring Raptor is at an advantage. At medium to high altitudes, the Typhoon is generally superior to the teen fighters in the within-visual range regime.

> ### PILOT ASSESSMENT
> **Best thing?** 'Its specific excess power [SEP].'
> **Worst thing?** 'How quickly you burn fuel when you are in reheat!'

ABOVE: Perhaps the most controversial Eurofighter customer was the Austrian Air Force. Allegations of bribery and corruption haunted the deal that resulted in a small force of high-performance interceptors.

DASSAULT RAFALE

Other than true stealth, the only disadvantage suffered by the Rafale – which has a good performance, an excellent defensive aids suite and a high level of sensor fusion – was the absence of a helmet-mounted display (HMD), but this has now been rectified; it is believed that Qatari Rafales are operating with it. The Rafale has a more advanced radar than other European and Russian fighters, and its weapons, the MICA and Meteor missiles, are less likely to be well known by 'threat' nations (the perennial Russian and Chinese bogeymen spring to mind) than the elderly and universal AMRAAM are, and are thus likely hard to counter. Though the long-ranged Meteor is held in very high esteem, it is worth noting that it has not been used in anger. The addition of HMD and the Meteor has made the already excellent Rafale even more potent and a strong contender for the title of the best multirole fighter in the world.

PILOT ASSESSMENT

Best things? 'It is very responsive and has very good situational awareness if you know how to manage all the screens and everything. It has a lot of capabilities: the omnirole stuff is very impressive; it can really switch extremely fast from air-to-ground to the air-to-air missions.'

Worst thing? 'The worst thing would be the noise; it's a pretty noisy aircraft. Like most of them, the environmental control system (the ECS) is pretty loud.'

HAL TEJAS

India's Tejas is the smallest and lightest of the modern fighters. Lampooned as a poor alternative to the Gripen, and criticised for its development issues, the Tejas has been an ambitious leap back into indigenous fighter design and manufacture following a long 'capability holiday' after Kurt Tank's Marut of the 1960s. The Tejas' distinctive wing features a shallow, swept inner section reminiscent of the Saab Viggen.

PILOT ASSESSMENT

Best things? 'The small size and low radar signature, coupled with a good sensor package, puts the Tejas in a good, advantageous spot with respect to bigger birds. The typical "first look, first kill" works very well for the Tejas in a fight, both in the beyond-visual and visual realms. The helmet-mounted display (HMD) is a very versatile piece of equipment and its system works well in a snap engagement, where the coupled missile "line of sight" modes allow the first shot to be good. The handling of the flight-control system is fabulous. The full-authority automatic low-speed recovery makes the aircraft truly carefree, more so than any other fighter in the world. This may be contested, but I'm willing to defend this position in a debate! Throw it around as much as you can – when she says no, she'll take over and recover the situation for you. The control and handling in high-gain tasks like aerial refuelling is superb; it will beat contemporaries or older birds in this area. It really makes you feel like a great pilot!'

Worst thing? 'Ironically, the size! It invariably tends to get compared to its bigger cousins in the business. The size essentially limits internal fuel, and hence the shorter legs as compared to others. However, if the focus is kept on the fact that it was intended as a light fighter, the fuel fraction is reasonable.'

SUKHOI SU-57 FELON

Russia's most advanced fighter incorporates a reduced-radar cross section compared with the Flanker. This is achieved through shaping, choice of materials and the inclusion of an internal weapons bay. Whether it offers a large enough jump in capability over the Su-35 (itself intended to plug the gap between legacy Flankers and Felons) to justify what is assumed to be a much higher cost remains to be seen. US levels of low radar observability, which require aircraft to have immaculately smooth outer surfaces, are not compatible with the Russian doctrines of 'rough and readiness'. Instead, the design has a greater emphasis on high performance, super-manoeuvrability and a counter-stealth capability (using unique, low-band wing-glove antennae). Working in conjunction with a stealthy unmanned combat air vehicle (UCAV), Sukhoi's S-70 Okhotnik-B, may provide a solution to some of its perceived shortcomings. The Su-57 has yet to see large-scale introduction.

PILOT ASSESSMENT

Best things? 'It has unimaginably good acceleration, extremely good handling at supersonic speeds and the ability to travel at high speeds in the lower stratosphere. It has impeccable handling characteristics.'
Worst things? Engine development issues, insecure funding, potential vulnerability to stealthier F-22.

SAAB GRIPEN E

The F-35 and Gripen E are digital-age fighters, but the later model, the Gripen E, has learned from the many software issues cursing the American aircraft. Saab, seemingly the only company able to run a combat-aircraft programme efficiently, has concentrated on making the Gripen E easy to upgrade in terms of software and hardware. This is a very serious advantage, as is the high degree of sensor fusion. If the Gripen E lives up to its promise, it will be an impressive machine. Larger and heavier than the Gripen, but without a proportionally larger increase in thrust-to-weight ratio, it may be a trifle more sluggish than its predecessor, however.

PILOT ASSESSMENT
(BY TEST PILOT JONAS JAKOBSSON)

Best thing? 'Situational awareness in the Gripen E is outstanding! All the way from the sensor suite (radar, IRST, missile-approach warner, radar warner etc.), the local fusion of sensor data in every Gripen and the global fusion of data shared within the tactical air unit (and C2) to the HMI with the elaborate symbology and wide-area display. This information chain and the situational awareness it creates is really the foundation that all fighting rests on.'
Worst thing? 'The worst aspect of the Gripen to me, personally, is that we are building such a fantastic and futuristic system, but it is all on the inside, so to speak. This makes it all a bit abstract and difficult to explain the full potential of the aircraft.'

SUKHOI SU-35 FLANKER

Among the most potent members of the originally Soviet Flanker series is the Su-35. The Su-35's supremacy may be challenged by the Chinese J-11D in some areas, notably the latter's AESA radar.

PILOT ASSESSMENT

Best things? 'The defensive electronics, which I can't talk about, are incredible. The Su-27's handling and performance were already world-beating, the '35 even more so.'

Worst thing? 'The man-machine interface is definitely behind the latest American jets. An AESA would be nice.'

MIKOYAN MIG-29 SERIES

Brutally fast and agile, the MiG-29 can hold its own against any fighter close in, but not for long, as most versions (especially the early ones) can carry only a small amount of fuel, and all have inefficient engines. The first fighter to have fleet-wide helmet-cued, thrust-vectored air-to-air missiles, it scared the bejesus out of Western military planners and many military pilots in the 1980s and early 1990s. MiG's political weakness, compared with the Sukhoi bureau, meant the type did not experience the same wealth of development enjoyed by the Flanker. The most advanced version, the MiG-35, has rectified many of the type's weaknesses but has arrived late with a price tag little different to the superior Su-35. Ludicrously overrated claims of the abilities of Russian warplanes are common, but Russian Aircraft Corporation MiG's claims of stealth and equivalence in combat effectiveness to the F-35 take the biscuit.

PILOT ASSESSMENT
(BY AIR MARSHAL HARISH MASAND)

Best thing? 'Its thrust-to-weight ratio, which was about 1.1:1 at take-off and came close to 1.3:1 at combat weight.'

Worst thing? 'Not enough gas. The upgraded versions now have more internal fuel as well as air-to-air refuelling.'

'Manufacturer Saab took great care to emphasise more sensible and boring qualities'

SAAB GRIPEN C/D

Spurning macho desires to make fighters ever faster, heavier and expensive to operate, with the Gripen, Swedish manufacturer Saab took great care to emphasise more sensible and boring qualities. This effort led to a maintainable fighter punching well above its light weight. Let's suppose you have a small-ish nation whose government does not have global dominance on its agenda. For such a nation, the key aim is deterrence, ensuring that any country wishing to invade or dominate you cannot easily do so. For such a nation, Gripen/Meteor might be the ultimate air defender, especially if you have a well-integrated air-defence system and dispersed bases. Never being far from the border or a base, fuel volume and even weapons-load don't matter so much, because you'll scoot back to your cave and re-arm and refuel. Having a big stick, however, is great, because you can defeat threats while keeping out of their missile range.

PILOT ASSESSMENT (BY LIEUTENANT MIKAEL GREV)

Best things? 'The large-display real-estate. Gripen C/D has a huge head-up-display [HUD] and three large, colour head-down displays [HDDs]. This gives ample opportunity to create a high-fidelity user interface for the pilot. The big HUD is good for dog-fighting (even though that's going away, as fun as it was) and the HDDs for BVR combat and basically all other mission types that are the chess-playing of today's air operations.'

Worst thing? 'The refuelling-probe length and position: it makes a mission component that should be easy and predictable an unnecessarily exciting part of the mission. Anecdote coming up! I've been told that when Gripen C/D was certified for air-refuelling, the refuelling expert pilot said something like "Gripen has probably the world's worst probe placement but compensates with the world's best flight control system." I concur with that statement. You can fly to the basket/drogue and stay easily within a metre or so of it, positioning your Gripen with almost centimetre precision with the stick, but when you approach it in the wake of the canopy, it will push it outwards. This means that you'll have to "go for it" and aim a bit on the outside of the drogue. This is not a good recipe for predictability. You do get good at it after a while and learn how to do it safely, but a longer probe wouldn't hurt.'

CAC/PAC JF-17 THUNDER

'Underrated, reliable and effective' – according to the pilot we spoke to – the JF-17 is a light fighter currently operated by Pakistan, Myanmar and Nigeria. While the Gripen utilises one Hornet engine to great effect, the JF-17 gets by with one MiG-29 engine. What the JF-17 lacks in all-out performance is offset by a major advantage: it is unaffected by US foreign policy decisions, as it is fitted with modern Chinese weapons and avionics (unlike the PAF's US-supplied F-16s). A generally conventional design, the most noteworthy feature of the aircraft is its diverterless supersonic inlet (DSI), which uses a 'bump' and forward-swept inlet to avoid the complexity of a variable-geometry intake.

PILOT ASSESSMENT

Best thing? 'Continuous upgrades of indigenous and Chinese weapons/electronic counter-measure suites, standoff capabilities of exceptional range, e.g. REK/IREK, CM-400 and C-802AK.'

Worst thing? 'The limited BVR load-out.'

LOCKHEED MARTIN F-35 LIGHTNING II

The F-35 was designed to excel in the attack role and to be used in conjunction with a specialised air-dominance fighter (the F-22), but its stealth and situational awareness make it a fearsome opponent in the beyond-visual-range fight. According to RAF Wing Commander Scott Williams, the BVR capabilities of the F-35 are 'second to none, really. First to see is first to shoot is first to kill. I recently heard a comment from someone that "fighting the F-35 is like going into a boxing match and your opponent doesn't even know you're in the ring yet!" I like that comment because our lethality is enhanced by being able to deliver the killer or knock-out blow to our opponents before they get enough awareness on what's going on to prepare or do something about it.' The F-35 is about stealth and situational awareness at the cost of the rather-more-exciting traditional fighter virtues.

PILOT ASSESSMENT

Best thing? 'How quickly and effectively the F-35 allows the pilot to make decisions – fusing sensor and other data from onboard and off-board sources to display what's out there and what's going on.'

Worst things? According to one pilot: 'I'd like a bit more fuel but what pilot doesn't?'; to another, the lack of 'feel' on the touchscreen, causing frequent inputting errors compared with a traditional cockpit, where buttons and toggles can be found without needing to be looked at, as well as the lack of a HUD.

LOCKHEED MARTIN F-22 RAPTOR

Way back in 1981, the new US president was a spritely 70-year-old called Ronald Reagan (back then, seventy was considered very old). Fervently anti-Soviet and anti-Communist, Reagan was happy to pay whatever was needed to destroy what he would later describe as the 'Evil Empire'. The Soviet Union then had two new fighters, the MiG-29 and Su-27, both of which were a serious challenge to the US 'teen' fighters' supremacy. Even more alarming were the new air-defence systems being developed. Soviet surface-to-air-missiles were becoming very effective and it appeared they would shred any conventional aircraft that came near. Combining the then new technology of radar stealth, supercruise (the ability to get to and sustain supersonic speed without recourse to reheat) and integrated avionics seemed to offer the solution. Little expense was spared to create this new super-fighter; it was very expensive, even for a US military programme. The result was a monster that, since entering service in 2005, has utterly dominated every other

potential fighter. It's so good that even the newer generation of Chinese and Russian fighters are considered inferior to the Raptor by most observers.

PILOT ASSESSMENT

Best thing? 'It's invincible – that's the bottom line. A phenomenal machine. It is unmatched.'

Worst things? Huge costs meant only a small number were made, and it was hugely maintenance-heavy and expensive to upgrade; it also had a less than stellar range, too few weapons to make the most of its advantages and no helmet display system (though one has been tested). It has also caused hypoxia and the mysterious 'raptor cough' respiratory condition in pilots.

MIKOYAN MIG-31

As a defender against bombers, the MiG-31 may well be the most potent interceptor in the world. The air defence of Russia drives you towards the MiG-31. You have to have a big, fast aircraft because you can't avoid the possibility of having to cover a fair distance at high speed to meet the threat. Being big means that a big sensor and long-range weapons are available, and both are likely to be needed. You may be less concerned about signature and platform manoeuvrability, because your ideal approach will be to stand back and hit bombers rather than engage fighters.

PILOT ASSESSMENT

Best things? 'The acceleration, the long-range radar and very long-range missiles. And it has been used to launch satellites. Massive top speed.'

Worst things? 'Lack of versatility, too clumsy for the dogfight. Poor visibility for the backseater. Situational awareness poor by modern standards. Perhaps the least stealthy fighter in the world.'

LOCKHEED MARTIN F-16 VIPER

Conceived as a light, simple and energetic dogfighter, the F-16 has matured into a complex middle-weight, multi-role aircraft but remains a nasty opponent in the 'knife fight in a phone booth' of close-in combat. Despite this, the US Air Force has largely used the type in the ground-attack role. The F-16 is the most common modern fighter and operated in air arms around the world. The latest versions have extremely capable sensors and modern cockpits.

PILOT ASSESSMENT

Best things? 'The view and HOTAS.'

Worst thing? 'The angle of the seat!'

CHENGDU J-10

Was it based on the Israeli Lavi fighter? The J-10 is undeniably based on CAC's own experiences with the cancelled J-9 project, which went through so many iterations during its long and protracted development, and surely the influence of the Israeli Lavi. But the Israeli contribution was more related to flight control system (FCS) development and integration, avionics and overall programme management than to the design of the fighter itself. Whatever the truth, it appears an excellent all-rounder, well equipped with modern sensors and weapons.

PILOT ASSESSMENT

Best things? Turning radius, low-speed performance and runway-length requirements. Well equipped with modern equipment.

Worst thing? High fuel consumption.

CHENGDU J-20 MIGHTY DRAGON

China is the second nation in the world to put an indigenous stealth fighter into operational service. With its extremely long-range anti-air weapons, this relatively stealthy platform could prove formidable. Unlike the Typhoon, the canard is not closely coupled to the wing, the main benefit from this arrangement being the carriage of significantly more fuel, coupled with the scope for a longer weapons bay, providing sufficient volume for a wide range of weapons. Stealth, supercruise and modern weapons mean the J-20 is likely to mature into an extremely capable aircraft, unique in role and configuration. Achieving this depends on the degree to which China can overcome its historical problems with engine development.

It is likely that the J-20 is less stealthy than the F-22 and F-35, and at least one F-35 pilot has stated that he doesn't believe the J-20 is low-observable in a meaningful sense.

PILOT ASSESSMENT

Best things? Likely to be fast, reasonably stealthy and well equipped, with very-long-range weapons and sensors.
Worst things? Probably not as stealthy as US equivalents; engine development issues.

MCDONNELL DOUGLAS F/A-18 C/D HORNET

The Hornet was probably the first 'modern' fighter in terms of cockpit and radar capabilities but is now in the twilight of its career. Developed from the Northrop YF-17, the losing competitor to the YF-16 in the USAF Lightweight Fighter contest, it was developed for the US Navy as a carrier-borne multi-role fighter. It has now been retired from the US Navy and Marine Corps and serves only in a handful of export customers' air arms as a land-based aircraft. Where other teen fighters enjoy unblemished air-to-air combat records, one US Navy Hornet was downed by an Iraqi MiG-25 in Operation Desert Storm of 1991. Famed for its 'turbo nose' – the ability to point its nose in the desired direction, however unlikely, with uncanny rapidity – the 'Bug' remains a deadly close-in fighter. Upgrades, including the addition of AESA, keep the F/A-18 relevant.

PILOT ASSESSMENT
(BY US MARINE CORPS PILOT LOUIS GUNDLACH)

Best thing? 'The Hornet was reliable. For the most part, the systems were almost always functioning well. The Marines working on the aircraft did a tremendous job keeping the Hornets flying and the systems working, often at austere locations. This is not only a testament to the Marines who worked tirelessly on the jets but to the reliability of the Hornet itself.'

Worst thing? 'The lack of fuel, at least around the carrier. I flew the Hornet for a long time in a land-based squadron. Fuel becomes a lot more critical when the runway is only open for 15 minutes every hour and a half or 2 hours. For carrier operations, the Hornet could have used more fuel.'

BOEING F/A-18E/F SUPER HORNET

The Super Hornet is akin to a luxury sports car without a big enough engine: it has all the 'bells and whistles' (a very powerful advanced radar, reduced radar cross section, an excellent cockpit, data-linking capability and good weapons) but lacks the grunt to make the most of its superb systems at higher speed and altitudes. The weapons carriage is also among the draggiest of any fighter. If the US Navy receives the new-generation BVR missile it wants, it is likely that the Super Hornet will be the first to receive it. Along with the Rafale, the Super Hornet, with its systems and weapons, is one of the best all-rounders.

NAVIGATOR'S ASSESSMENT

Best things? 'The fact that all the systems work well together and that it's a very reliable aircraft maintenance-wise. No conventional airplane can point its nose around at slow airspeed like a Rhino (or Hornet).'

Worst thing? 'The drag from those dumb, goofy crooked pylons.'

'The Super Hornet is akin to a luxury sports car without a big enough engine: it has all the "bells and whistles" but lacks the grunt to make the most of its superb systems at higher speed and altitudes'

MCDONNELL DOUGLAS
F-15 EAGLE

Few fighters have been so highly regarded for so long. Despite first flying as far back as 1972, the best-equipped Eagle remains among the deadliest fighters in the world. Its kill-loss ratio of 104:0 is without precedent. The next variant for USAF, the F-15EX Eagle II, will be the best-equipped fighter in the world.

PILOT'S ASSESSMENT (BY PAUL WOODFORD)

Best things? 'The excellent radar, the ergonomically designed weapons controls on the stick and throttles, the weapons displays on the HUD (and later on the helmet visor), the high seating position and unrestricted cockpit visibility. The Eagle is a direct descendant of the F-86 Sabre, another outstanding air-to-air fighter with great cockpit visibility. Did I mention the cockpit's also quite roomy? Having flown in Century Series fighters, I'm here to tell you that's a big deal.'

Worst thing? 'Its huge size makes it easy to spot with eyes; its huge radar cross section makes it easy to spot on radar. Though once a monster in the merge, it is surpassed in manoeuvrability by a new generation.'

MODERN FIGHTERS COMPARED

Disclaimer: These figures are illustrative estimates based on public domain sources and should not be used for air combat planning.

MAXIMUM SPEED

Top speeds are limited by several factors, notably air-intake design, materials and aerodynamics. The MiG-31 is a specialised interceptor based on the earlier MiG-25 that prioritises speed, at the cost of agility, and is in class of its own. Though top speeds are rarely reached, a supersonic dash can mean the difference between life and death in a combat situation. Importantly, a missile launched from a faster aircraft will travel further.

Light

JF-17	1.6
Gripen C	1.8
Tejas	1.8

Medium

MiG-35	2.2
F-16C	2.0
Typhoon	2.0
J-10C	1.85
Gripen E	1.8
Rafale C	1.8
F-35	1.6
F/A-18E/F	1.6

Heavy

MiG-31	Wartime limit	2.83
J-20		2.35
Su-35		2.35
F-15C		2.3
F-22		2.25
Su-57		2.0+
MiG-31		1.5	Peacetime limit

FRONTAL RADAR CROSS SECTION (X-BAND)

The first fighter to be detected is the most vulnerable, thus a low frontal radar cross section is an advantage. Hard figures do not exist in the public domain, but this diagram illustrates the aircraft with the biggest likely advantages and disadvantages in this area. The F-22 and F-35 were designed for offensive operations, where stealth is more important; the MiG-31 is designed for a defensive role against bombers, where stealth is less important.

Light

Gripen C	5
JF-17	5
Tejas	4.5

Medium

MiG-35	15
F-16C	8
Gripen E	5
J-10C	5
F/A-18E/F	4.5
Typhoon	4
Rafale C	3.9
F-35	· 0.05

Heavy

MiG-31	38
Su-35	25
F-15C	21
Su-57	3
J-20	2
F-22	· 0.03

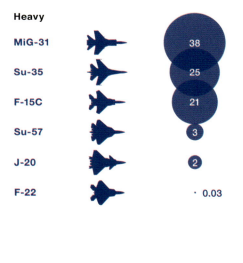

BEYOND-VISUAL-RANGE LOAD-OUTS

Until the 2010s, ultra-long-range air-to-air missiles were only carried by the F-14 and MiG-31. Recently all major nations have developed, or are developing, such weapons. There has been little beyond-visual-range air combat, and nothing approaching the quoted figures for modern weapons. It remains to be seen if such warfare is possible in the real world or whether it is more in the realm of deterrence. This depends heavily on the rules of engagement put in place by air arms for air combat, and the degree to which positive identification is required before weapons can be released.

The Gripen E's given load-out would likely be smaller in most real-world missions, though it could be adopted in emergencies.

> 'Until the 2010s, ultra-long-range air-to-air missiles were only carried by the F-14 and MiG-31. Recently all major nations have developed, or are developing, such weapons'

FUEL FRACTION

The fuel fraction of the aircraft at typical combat weight, without external tanks, is a key measure of its likely combat persistence. External tanks increase supersonic drag and impose carriage limits. Though expensive and logistics-heavy, air-to-air refuelling means fighters can extend their range in flight and their combat persistence. The Flanker is ultra-agile at low operating weights, but with maximum fuel and weapons it is rather cumbersome.

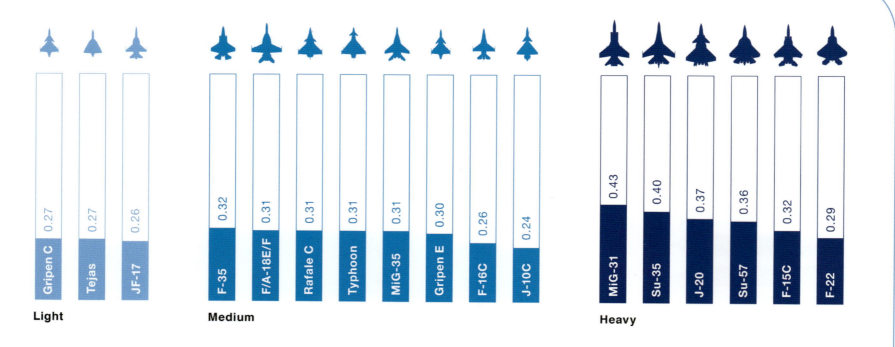

Light
- Gripen C 0.27
- Tejas 0.27
- JF-17 0.26

Medium
- F-35 0.32
- F/A-18E/F 0.31
- Rafale C 0.31
- Typhoon 0.31
- MiG-35 0.31
- Gripen E 0.30
- F-16C 0.26
- J-10C 0.24

Heavy
- MiG-31 0.43
- Su-35 0.40
- J-20 0.37
- Su-57 0.36
- F-15C 0.32
- F-22 0.29

Range and number of weapons

● Light ● Medium ● Heavy

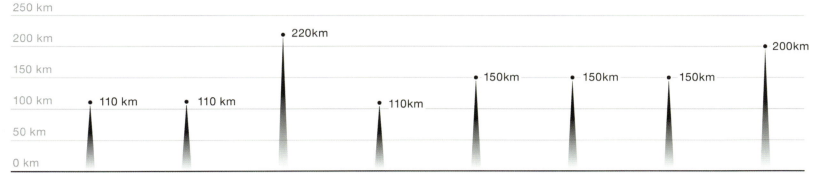

Tejas	JF-17	Gripen C	MiG-35	F-35	F-16C	F/A-18E/F	J-10C
4 x R-77-1	4 x SD-10A	4 x Meteor	8 x R-77-1	4 x AIM-120D	4 x AIM-120D	10 x AIM-120D	4 x PL-15
	2 x WVR				2 x WVR	2 x WVR	2 x WVR

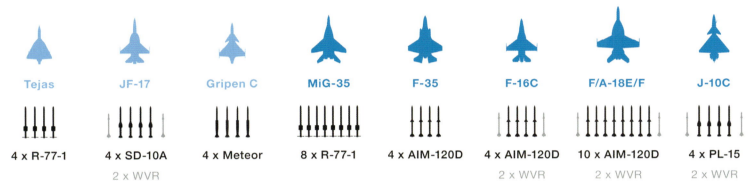

Rafale C	Typhoon	Gripen E	Su-57	F-22	F-15C	J-20	Su-35	MiG-31
4 x Meteor	4 x Meteor	7 x Meteor	6 x R-77-1	6 x AIM-120D	8 x AIM-120D	4 x PL-15	8 x R-77-1	6 x R-37M
4 x MICA	4 x WVR		2 x R-74	2 x AIM-9X		2 x WVR	2 x R-37M	2 x R-74

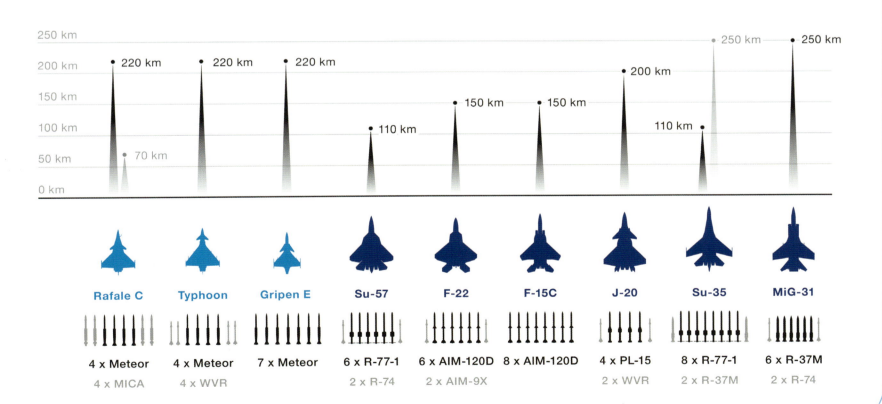

RADAR RANGE

'First look' is required for 'first kill' and a large powerful radar helps. The US currently dominates the radar technology field but Europe and China are catching up. A larger nose can carry a bigger radar, which, at a given technology level, will increase the detection range.

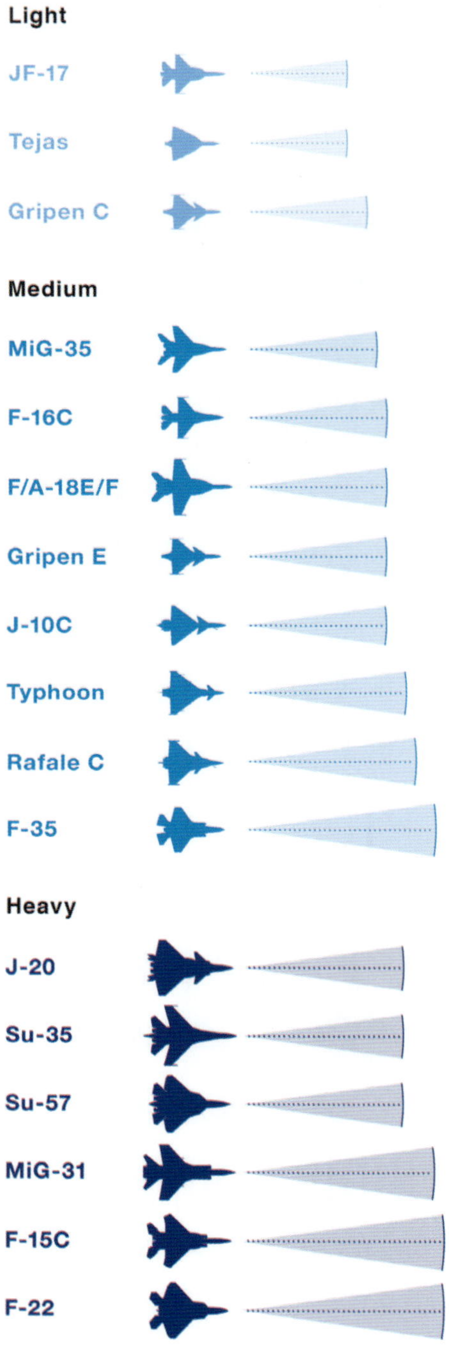

Light

JF-17

Tejas

Gripen C

Medium

MiG-35

F-16C

F/A-18E/F

Gripen E

J-10C

Typhoon

Rafale C

F-35

Heavy

J-20

Su-35

Su-57

MiG-31

F-15C

F-22

THRUST-TO-WEIGHT RATIO

Thrust-to-weight at combat weights is revealing: the more spritely performer will have greater ability to manoeuvre and to then regain energy – an enormous advantage if used carefully. Though only the Typhoon's wartime levels are shown, all fighters have an emergency setting that boosts power at the cost of engine life.

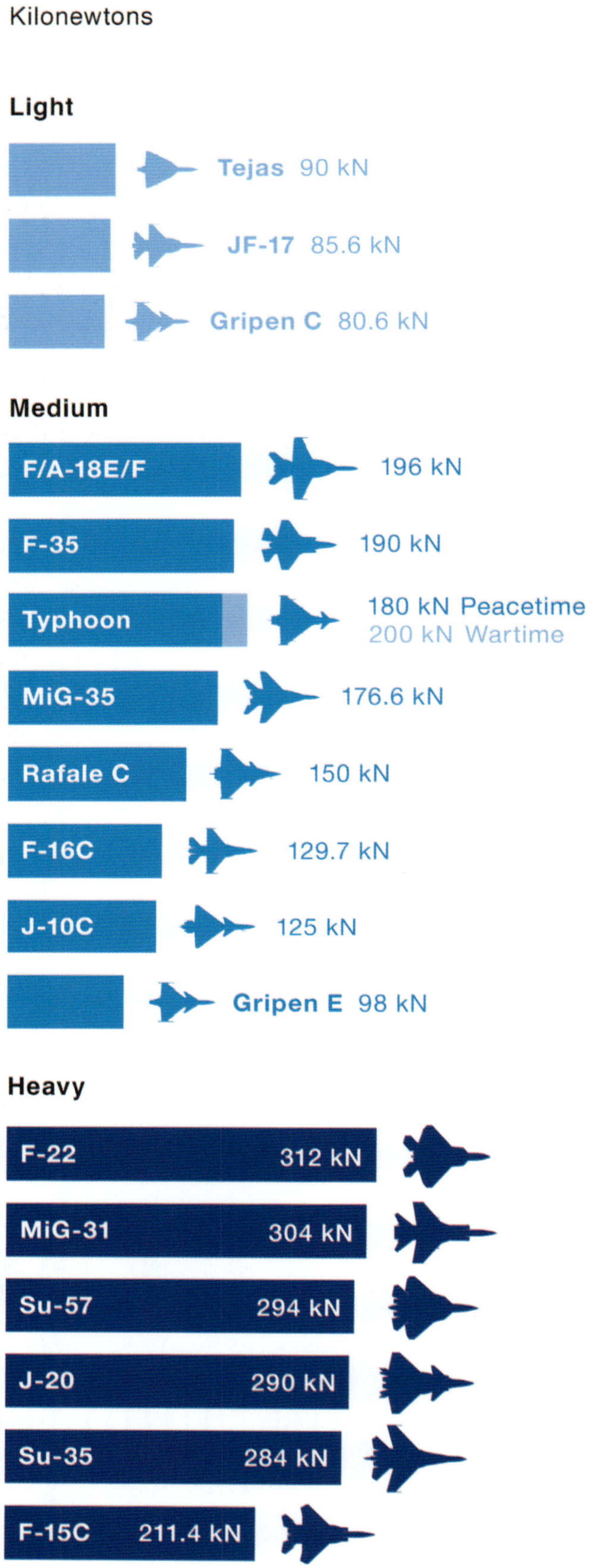

Kilonewtons

Light

Tejas 90 kN

JF-17 85.6 kN

Gripen C 80.6 kN

Medium

F/A-18E/F 196 kN

F-35 190 kN

Typhoon 180 kN Peacetime / 200 kN Wartime

MiG-35 176.6 kN

Rafale C 150 kN

F-16C 129.7 kN

J-10C 125 kN

Gripen E 98 kN

Heavy

F-22 312 kN

MiG-31 304 kN

Su-57 294 kN

J-20 290 kN

Su-35 284 kN

F-15C 211.4 kN

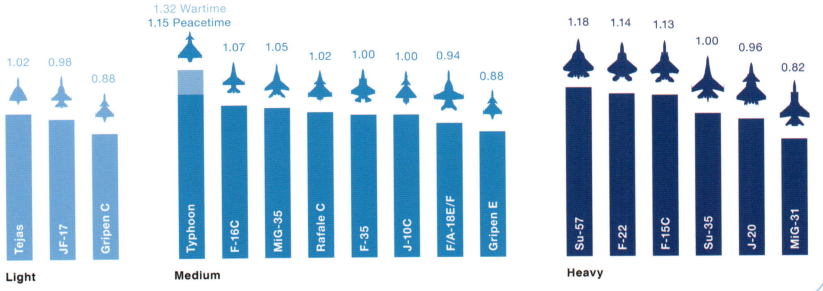

Light

Tejas 1.02
JF-17 0.98
Gripen C 0.88

Medium

Typhoon 1.32 Wartime / 1.15 Peacetime
F-16C 1.07
MiG-35 1.05
Rafale C 1.02
F-35 1.00
J-10C 1.00
F/A-18E/F 0.94
Gripen E 0.88

Heavy

Su-57 1.18
F-22 1.14
F-15C 1.13
Su-35 1.00
J-20 0.96
MiG-31 0.82

VANQUISHING
VIPERS!

Whenever we interview a modern fighter pilot, one subject that always comes up is how his or her fighter compares to the F-16 in a close-in fight. I collated the answers I received for a snapshot of how the pilots of other types rate the formidable Viper.

MIRAGE 2000 VS F-16

Ian Black: 'I must have flown against the F-14, F-15, F-16, F-18, Tornado F3, F-8 Crusader and the F-104 Starfighter in combat. The older generation didn't stand a chance, but the F-16 Block 50 was very good. I got beaten by an F-16 by fighting him like a Mirage and learned a painful lesson. DACT was interesting in the M2000 – if your opponent was new to fighting a delta, it could make his eyes water! At the merge the initial 9-G-plus turn was gut-churning; despite having a single engine it could still reach heights other fighters, like the F-16, couldn't. In my opinion, it also possessed a far more sophisticated fly-by-wire system – it was in effect limitless. I managed to put a Mirage 2000 into the vertical whilst being chased and held the manoeuvre a few seconds too long – when I looked into my HUD I was in the pure vertical at 60 knots and decelerating! As we hit zero the aircraft began to slide backwards and the 'burner blew out. My heart rate increased. As the aircraft went beyond its design envelope, the nose simply flopped over, pointing earthwards – but with a few small turns the airspeed picked up. As I hit 200 knots I simply flew the aircraft back to straight and level. I admit that my opponent did shoot me down, but he did say it looked spectacular. This sort of carefree handling gave pilots huge confidence in the aircraft.'

GRIPEN VS F-16

Would you be confident facing an F-16? And how do the types compare?

Lieutenant Mikael Grev: 'Absolutely. I can't think of anything the F-16 would be better at, if we don't count ease of refuelling (the F-16 is refuelled with a boom and the boom operator does much of the job). Of course, there are a lot of details and circumstances here, but generally the Gripen is a step or two ahead, especially in my favourite areas. I really like pilot user interface and large screens, and the F-16 is lacking a bit in that area so maybe I'm a bit biased. I do like the F16's side-stick, though! It didn't take many minutes to get used to the stiffer stick, and it's more ergonomic for the pilot in high-Gs (and probably for long missions) to have it on the side. Flying in close formation with another fighter was almost as easy as with the Gripen.

'The F-16s are a lot like the Gripens but you can claw yourself closer and closer to their behind, if that is your goal.'

F/A-18C HORNET VS F-16

What is the best way to fight an F-16? And the worst?

Louis Gundlach: 'Throughout my career, I flew against F-16s many times, and in my opinion it was the hardest of the fourth-generation fighters to beat. It was small, had a lot of thrust and a very impressive 9-G turn. The F-16 had a turn-rate advantage and much better thrust-to-weight when compared with the F/A-18C, but the F/A-18C had a better turn-radius and could fly at a higher angle of attack (AOA) than the F-16. The best way to fight an F-16 is in a one-circle fight, usually in the vertical, getting the Hornet's nose on first to try and get an early shot, whether with a missile or the gun. The key would be to get the F-16 reacting to the Hornet, bleeding energy and getting slow. At slow airspeeds, the F/A-18's AOA advantage meant I could point my nose easily and get a shot. The worst way to fight against an F-16 would be a two-circle fight on the horizon. The F-16's 9-G turn and superior thrust-to-weight would give him a better turn-rate and the F-16 would out-turn the Hornet. If an F/A-18 tried to match the F-16 turn-rate, the Hornet would get bleed energy and its turn-rate would continue to be less than the F-16's.

'Like all fighters, most of a fighter plane's ability to fight is dependent on the skill of the pilot. The F-16's performance, much like the Hornet's, would suffer if it was carrying external stores. A slick Viper flown by an experienced pilot was a beast and always a tough fight. There was an air force reserve squadron out of Luke that was full of experienced pilots – all of them had at least a thousand hours in the Viper. They always flew slick Vipers and they were a tough fight for an F/A-18C, which always had at least one external tank and two pylons. This reserve squadron also went on that Key West Det. From what I saw and experienced, in a purely visual fight a slick Hornet was better than a slick Viper. I rate the

OPPOSITE: Despite its age, the F-16 remains an extremely challenging dogfight opponent. The Viper, as it is universally known, is a small aircraft, making it harder to see, and was designed with a high thrust-to-weight ratio, ensuring it retains energy well in a 'furball'. It featured the first use of relaxed stability with fly-by-wire controls, which are today included in all modern fighters.

ABOVE: The most colourful F-16s are those simulating hostile aircraft for USAF Aggressor units. The pilots of the USAF Aggressor F-16s are considered by some the best in the world, spending a great deal of time in mock combat with leading fighters, including the almost omnipotent F-22 Raptor.

F-16 pilots from that reserve squadron in Luke as the best I ever fought, and in the visual arena the Hornet more than held its own.'

RAFALE VS F-16

Pierre-Henri 'Até Chuet: 'The F-16 is pretty cool. The Typhoon is a joke, very easy to shoot. The F-16 was a good surprise, actually – I found it to be a pretty good aircraft. I think the most challenging was the F-16; it's a pretty small jet so it's easy to lose sight of it. So, I think that was the big one.'

TYPHOON VS F-16

Squadron Leader Roger Cruickshank: 'The Typhoon is a superior fighter within visual range, though we must always remember that we are not fighting the aircraft but the pilot.'

Of the aircraft you trained against, which was the hardest opponent and why?

RC: 'I fought a TOPGUN instructor out of Nellis Air Force Base and he was in an F-16. I was not very experienced at the time, though I managed to defeat him. He did, however, make it very difficult!'

Air Commodore Paul Godfrey: 'I have flown basic fighter manoeuvres (BFM – 'dogfighting') against a Viper several times in a Typhoon. The Typhoon generally comes off best due to its raw power. Even at slow speed you can plug the burners in and go up. The F-16 was sometimes a little tricky to get speed back if you had dumped it in a manoeuvre. However, a well-fought F-16 is definitely a formidable opponent. One additional Typhoon advantage is the world-class anti-G system, which allows you to pressure-breathe, while also using an anti-G suit and anti-G pressure vest which both

fighter pilot was the F-16 of the Pakistan Air Force (for obvious reasons). So the excitement of facing an F-16, even in a mock combat, was unbelievable. The weight of the mission was overbearing! Perhaps that's what makes it special. As the combat commenced, we manoeuvred for our lives and in very little time the situation was in our favour. The desperate calls from the F-16 – *"Flare! Flare! Flare!"* – are very distinctly audible in my ears even today. From that day, the anxiety that prevailed over facing an F-16 in combat was gone for ever – vanished! It was clear what the outcome would be.

'Another mission that stands out is a group combat mission that was pitching a Su-30 and an MiG-21 Bison against three F-16s. As luck would have it, the Bison didn't get airborne and now the game was one Su-30 vs three F-16s in a beyond-visual-range scenario. Again, we pushed the envelope, manoeuvred between 3,000 feet to 32,000 feet, pulling up to 8-G, turning, tumbling, firing and escaping missiles in a simulated engagement. The crew coordination between us in the cockpit and the fighter controller on the ground was the best that I have ever seen! The results in a mock combat are always contentious but with ACMI they are more reliable. End score: one F-16 claimed without loss. When we got out of the cockpit we were thoroughly drenched in sweat and tired from the continuous high-G manoeuvring, but all smiles for the ecstasy that we had just experienced.'

F-15C VS F-16

Major Shari Williams (Rtd): 'Assuming equal pilots (meaning both have the same air-to-air experience and recency of experience), the F-16 is a more efficient-turning plane. It enjoys a slight advantage in sustained-turn ability, whereas the Eagle has a slight advantage in instantaneous-turn ability. The turn circles are almost identical. Depending on configurations, the thrust-to-weight ratio is all pretty close to equal. So, how did I fight an F-16? First, I always assumed the pilot was awesome. Assuming we meet 180 degrees out, with our speeds where we want them, and no one has an angular advantage, I would elect to take the fight single-circle (the tactical scenario may not favour this in a full-up air battle). My goal is to get slow and use my ability to fly at higher AOA/slower speeds than the F-16 can. The F-16 has decent AOA capability but the fly-by-wire system is limited in speed of movement of the controls as it approaches its AOA limit. The F-15 has no such limits. I usually had more air-to-air experience than the vast majority of F-16 pilots and usually had little trouble neutralising and then killing them in close. Like all victories, it comes down to flying your particular aircraft at the extremes and doing it more efficiently and precisely than the

inflate to help you stay conscious at high G. F-16s, whilst they have an anti-G suit and a system called Combat Edge (an inflatable vest), you are working very hard to stay awake.'

MIG-29 VS F-16

Air Marshal Harish Masand: 'In a modern MiG-29 like the upgraded one or the M version, and trained well, I feel the pilot should be supremely confident against the modern F-16.'

SU-30 VS F-16

Captain Anurag Sharma: 'I have had many memorable missions over the years but the few that stand out include a DACT with F-16 Block 52 of the Republic of Singapore Air Force. The strongest adversary that we could possibly face in our life as a

other pilot. That being said, an F-16 can win a single-circle fight if the adversary is not on their game. It can also lose a two-circle fight if the pilot is not proficient at it.

'Let me add this. Air-to-air combat is incredibly fluid; it changes very fast. So even though an F-16 may have a better sustained-turn-rate than an F-15C, if through my intercept I can achieve 30 or more degrees of lead-turn, I will happily go two-circle. And that is the goal, to merge with an advantage; that way, any enemy advantage is minimised and maybe even negated and a quick kill follows. That is the goal!

'In my 2,000-plus hours, I fought the Viper a lot; I have flown against many Weapons School grads and against average pilots. In most all cases, I did really well. For any fighter pilot, it is about controlling the fight and forcing the fight that favours your aircraft. Because most F-16 units don't do much air-to-air (adversary tactics folks being the exception); their experience, especially their currency, was often spotty at best. So, was I confident? Always. Did I do well? Usually. But everyone has bad days and good days. That is why there are no absolutes in air-air combat.'

JF-17 VS F-16

Which threat aircraft is most challenging and why?

'The Su-30 is definitely the most difficult aircraft in terms of current Indian Air Force inventory but we regularly fly against the F-16 and, more importantly, AMRAAM, so Adder and Alamo seem less worrisome.'

How comfortable and ergonomic is the JF-17 cockpit?

'The cockpit is one of the most digitised I've flown to date. Even the F-16's cockpit pales in comparison to the Thunder's cockpit layout.'

In a WVR fight would you rather be in an F-16 or JF-17?

'F-16 for the initial 180-degree turn, then the Thunder all the way. The JF-17 with PL-10 mod (currently in the pipeline) will trump the F-16 with the AIM-9M any day of the week, but currently, on brute performance, the F-16 has the edge.'*

MIG-21 VS F-16

Group Captain M. J. A. Vinod: 'If you asked me whether I'd feel more confident going against a modern F-16 or MiG-29 in a

---

* Pilot name withheld by request.

MiG-21, I'd tell you that is not an apples-to-apples comparison. Modern-day fighters have systems assisting you – superior radar, helmet-mounted sighting systems, great radar warning receivers, counter-missile systems, electronic warfare systems like the self-protection jammers, etc. The older-version MiG-21s had none of these, so they are clearly out of the fray. The MiG-21 Bison is the most modern MiG-21, and it is formidable – the only downside being the limited endurance. Eventually it is the man–machine combo that makes or breaks an air combat.'

F/A-18F SUPER HORNET VS F-16

'A fight I had in 2009 is a good example of WVR against an F-16. We were on detachment to Key West (the real fighter-pilot heaven) and we were fighting against Air National Guard pilots in their F-16Cs. I was a senior WSO paired with a new pilot who had been with the squadron less than a year but who had flown combat with us in Afghanistan, all air-to-ground. The F-16 we were fighting was armed with AIM-9M and flown by an Air Guard lieutenant colonel. He was experienced, but we had the JHMCS helmet and AIM-9X. Often our Navy tactics were based on observing what the bandit does and then executing a game plan based on that, but I instead scripted our first two moves so that New Guy would have a very clear mental picture of what to do and be able to execute it.

'Our game plan was that at the first head-to-head pass we would immediately go down in a split S, regardless of what the F-16 did. Air Force doctrine is to not highlight yourself in a cold, blue sky against a guy with an advanced heat seeker. So I assumed that he would come down with us, then when we met him again, we would go down again regardless. We would essentially be in a one-circle fight going downhill with gravity helping us stay fast. The idea was that we would meet at the hard deck on the third pass and both aircraft would have a ton of speed, and then we'd pull the surprise. If my assumptions were wrong, it could get ugly fast.

'We met out over the water in the mid-20s [20,000-foot altitude range] and the fight's on. New Guy immediately went down, and sure enough the F-16 came down with us. The one-circle geometry kept us inside his 9M forward quarter minimum range. A second head-to-head pass and we immediately went down again. . . And the F-16 came down again. We now have a third merge just above the hard deck and both of us have a ton of energy. Now here comes the surprise: nothing slows down like a Rhino with those big, goofy, crooked pylons on the wings. And no airplane without vectored thrust can point its nose around at slow airspeed like a Rhino (or a Hornet, for that matter).

'For a while, our squadron had a jet with no tanks on it that we were using for air-show practice'

'At the merge, the F-16 started a high-G turn but, with all the speed he had, he was cutting a pretty wide circle above the hard deck. When he looked over his left shoulder, he saw our jet pointing at him, seeming to almost hang in the air. New Guy had the F-16 in the HUD with a screaming AIM-9X tone.

'You see, circa 2009, nothing in the regular inventory could slow down and point its nose like a Rhino (except maybe a helicopter). Now, we couldn't come across the circle and chase the F-16 down; we were at very low airspeed and not really going anywhere at this point. But due to the phenomenal flight-control computers banging around all twenty-four flight-control surfaces multiple times per second, we were able to keep flying and pointing our nose at him. The colonel knocked it off and New Guy had a sweet HUD tape of him nose-on to an F-16 in plain view. Of course, if there had been more than one F-16, floundering around at low airspeed would have made us a tasty target for his wingman. But on this day – no wingman, no problem. The moral of the story is to be careful getting into visual range of a Rhino; he can't run away, so he'll stay and fight because he has to.

For a while, our squadron had a jet with no tanks on it that we were using for air-show practice. We'd take that thing out and BFM in it when we weren't practising. That thing would eat even the F-15s and F-16s for lunch.'

Which set-ups and altitudes would the F/A-18F favour?
'I'd like to be lower in the twenties or teens, looking up at the F-16. The Rhino doesn't do great up high (unless it's clean, with no pylons), and of course, looking up at him makes things the easiest for my sensors (including my eyeballs) and hardest for his.

'At range, shoot and let the AMRAAM do its thing. If bandits were blowing up and their formations falling apart, we'd go to the merge and press our advantage. If that wasn't the case, we could fall back, regroup, and try again. Once you get to BFM, the Rhino will take most adversaries one circle.'*

---

* Pilot name withheld by request.

IMAGE CREDITS

Afbase.com: 146 (top)

Airwolfhound: 15

Alamy: 129

Armytech.net: 147 (middle)

Talgat Ashimov: 179 (bottom)

BAE Systems: 8, 9, 19, 24 (right), 34 (via The Aviation Historian), 64, 77 (top; via Wyvern Images), 78 (bottom; via Wyvern Images), 111–5, 118–9, 122 (bottom), 161

Bangladesh Air Force: 190 (bottom)

Ian Black: 59, 121, 123, 140–43, 146 (bottom), 172–4

Boeing–Sikorsky: 137 (bottom)

Pierre-Henri Chuet: 181 (top)

Commonwealth of Australia – Department of Defence/Sgt David Gibbs: 196 (left)

Convair: 73 (top; via Hansa Images), 74 (top; via Hansa Images), 138 (bottom right)

Dassault Aviation S.A.: 147 (bottom), 159 (right), 180, 181 (bottom), 182, 184–5, 188 (left)

Stephan de Bruijn: 150

DIA/public domain/Richard E. Terry: 136 (top left)

EADS: 158, 159 (left)

EADS/Airbus: 155 (left)

Eurofighter GmBH: 148 (top left)

Eurofighter/G. Lee: 179 (top)

Barry Farquharson: 58

Andrew Green: 57 (top left)

Grumman: 138 (bottom left), 139 (top)

Hansa Images: 50 (bottom), 71 (bottom), 94 (top)

Jamie Hunter/Aviacom Ltd: 208

IanC66/Shutterstock.com: 16

Indian Air Force: 94 (bottom), 96 (top), 176, 178

Inkworm/Chris Sandham-Bailey: 61, 80, 86, 152, 165

Islamic Republic of Iran Air Force: 151, 166–9

Penny Klein: 102

Vitaly Kuzmin: 147 (top)

G. Lee/Plane Focus/Eurofighter: 187

Lockheed: 135 (bottom)

Lockheed Aeronautical Systems: 134

Lockheed Martin: 72 (bottom), 74 (bottom), 75 (bottom), 148 (bottom right)

Luftwaffe: 11 (bottom), 120 (via Hansa Images), 122 (top; via Hansa Images)

Marina Militare: 192 (right)

Martin-Baker: 69, 76, 83

McDonnell Douglas Aircraft: 135 (top), 136 (top right)

Venkat Mengud: 188 (right)

MiG: 46, 57 (top right), 105 (via Hansa Images)

NASA Glenn Image Gallery Archives: 139 (bottom right)

NASA/public domain: 157 (right)

National Museum of the USAF: 17, 145

Kirill Naumenko: 124

Northrop Grumman: 73 (bottom)

Pakistan Air Force: 192 (left)

Bud Parke/Grumman: 139

Pikabu.com: 146 (middle)

PLAAF: 194 (bottom), 195

Abdullah Bagheri Raad via Babak Taghvaee: 170

Tom Rosquin: 107 (via B. C. Thomas)

Royal Danish Air Force: 126 (top)

Oren Rozen: 29 (top)

Russian Air Force: 194 (top left)

Ryan Aeronautical: 154 (via SDASM Archive)

Saab: 162, 164 (top), 189 (right), 191

Abdollah Sharifi-Raad: 171

Anurag Sharma: 177

Valeri Shatrov: 89–91

David Shultz: 28

Soviet Navy: 156 (top), 156 (bottom; via Hansa Images)

Sukhoi: 11 (top), 52 (bottom right), 53 (bottom; via Hansa Images), 189 (left), 190 (top)

Keith Tarrier/Shutterstock.com: 104

Teasel Studio: 13 (bottom), 14 (top), 18, 20, 30 (top), 35, 36 (top) , 46, 47–9, 50 (top), 52 (top), 53 (top), 54–6, 62, 65, 75 (top), 78 (top), 81 (bottom), 82, 84, 96 (bottom), 116, 127, 160 (bottom)

The Aviation Historian/Nick Stroud: 38

B. C. Thomas: 109 (top)

US DoD: 77 (bottom) , 136 (top)

US Navy: 27 (left), 31 (via Wyvern Images), 43, 45 (bottom), 68 (right; public domain), 125, 136, 144, 148 (bottom left), 196 (right)

USAAF: 26, 27 (right), 33, 40, 45 (bottom), 63

USAF: 10, 66, 68 (left), 70, 72 (via Hansa Images), 98–101, 103, 108, 109 (bottom), 128, 130–132, 155 (right), 156 (top left), 183, 186, 194 (top right), 197, 199, 200, 203–4, 206

US Army: 157 (left)

USMC: 149, 160 (top left)

Alan Wilson: 22

Wyvern Images: 23 (public domain), 25 (public domain), 29 (bottom), 32 (public domain), 37 (public domain), 39, 41, 42 (top; public domain), 45 (top; public domain), 51, 52 (bottom left), 60 (public domain), 110

Zhihu.com: 79 (top)

A NOTE ON THE EDITOR

Joe Coles has been writing about aviation for over twenty years. In this time he has been the editor of a bestselling aviation periodical published around the world, and contributed journalistic and historical articles to many leading publications including DK's *The Aircraft Book*, which has been translated into several languages. He created Hush-Kit, a widely respected and deeply irreverent online magazine. He lives in Bristol.

hushkit.net

unbound

Unbound is the world's first crowdfunding publisher, established in 2011.

We believe that wonderful things can happen when you clear a path for people who share a passion. That's why we've built a platform that brings together readers and authors to crowdfund books they believe in – and give fresh ideas that don't fit the traditional mould the chance they deserve.

This book is in your hands because readers made it possible. Everyone who pledged their support is listed below. Join them by visiting unbound.com and supporting a book today.

Nicholas Aarons	Jason Andrade	Simone Ba	Antony Barton
Avioniq AB	Bernard Angell	Peter Bäckström	Conor Barton
Walter Abeson	David Antol	Guillaume Bacry	Thomas Bastiani
Robert Abram	Brian Anzaldua	Mark Baehr	Sagar Basutkar
Dale Abramson	Connor Appleton	David Baez-Kasolis	Edward Bateman
Chris Adams	David Aramant	Mark Baffa	Peter Batty
Graeme Adamson	M R Arden	Seth Bagby	Christian Bauer
Timothy Adkins	Jason Arends	Alex Bailey	Matthew Bayer
Jim Aitken	Greg Armstrong	Robert Baird	Gary Beale
Alan Akeroyd	Wade Armstrong	Nils Baker	Trevor Beattie
Ben Akrigg	Juan Artigas	Krishna Balaji Moorthy	Chris Beaumont
David Aldworth	Ron Artigues	Adam Bancroft	Jan Becherer
David Alexander	David Ascroft	David Baranek	David Beck
Alix	Glenn Ashcroft	Dean Baratta	Marc Bedau
Cam Allan	Richard "Ash" Ashcroft	Dale Barber	Chris Beddoes
David Allan	Ben Askin	Carl Barbour	Alex Beeman
Ashley Allen	Jonathan Astley	Alastair Barker	Rolf Behrens
Paul Allen	Kevin M. Atamanchuk	Ed Barnard	Rory Bell
Stephen Allport	Denny Atkin	Steven Barnes	Christopher Belyea
Flt Lt Dan Allum 30th	James M. Atkins	Susan Barnes	Jesus Maria Martin Benavente
Per Alriksson	T. Atkins	Nelson Barnhouse	Paul Bennett
David Ambrose	Eric Atwell	Carl Baron	William Bennett
Joseph Ambrose	Mark Aubry	Dom Barr	Michael Bentley
Alan G. Ampolsk	Jacob Augustyn	Stephen Barraclough	Philip Bentley
Bruce Andersen	Charles Aunave	Colin Barrett	Dominic and Alexis (Bethencourt-)
Richard M Anderson	Stewart Ayrey	Paul Barrett	Smith
Magnus Andersson	Zaharias Azhar	John Barrow	Klas Bergenheim
Josh Andjelkovic	George B.Kinghorn	Matt Barry	Jonathan Berger

David Beringer
Simon Berry
Steven Bienkowski
Chris Biggers
Chris Binner
Ellice Birnie
Øyvind Bjørkli
Eileen Bjorkman
Mike Black
Edward Blackburn
Jason Blackstone
Dan Blase
James Blatch
Graham Blenkin
Christopher Blom
Alexander K. Blu
Andy Blundell
Carsten Böhme
Mark Bomhof
Matthew Bone
Andrew Booth
Simon M Booth
Frank Bosman
Filippo Botti
David Bottomley
Simon Boughton
Marc Boulton
Jacco Bourgonje
Florent Bourquard
Dan Bouvier
Kevin Bovis
Chris Bowen
Adam Bower
David Boyd
Tristan Boyd
Dan Boyle
Andy Bradshaw
Simon Bradshaw
Toby Brampton
Claus Just Brandstrup
Donal Brannigan
James Branthwaite
Peter Branthwaite
Ray Brase
Dan Bratton

Harry Braviner
Terence Bray
David Braybrooke
Richard Brayshaw
Nicholas Brazel
Anthony Bright
John Brimlow
Alkie Brindley
Neil Briscoe
Pete Broadbere
Adam Brockie
Joakim Brolin
Martin Bromiley OBE
Justin Bronk
Jonathan Brossart
Aaron Brown
Cory Brown
Jesse Brown
Joseph Brown
Neil Brown
Ryon Brown
William Brown
Brian Browne
David Browne
Alasdair Bruce
Alice Bruderer
Ian Brumby
Philippe Bruneau
Maxwell Brusky
Vincent Bryson
Matt Buchanan
Matt Buchanan
Matthew Bucher
Timothy Bucklin
Marcus Bufton
Elias Bühler
Rob Bullen
Federico Burchianti
Luke Burden
Andrew Burgess
Neil Burkinshaw
Geoffrey Burks
Rob Burnett
Chris Burns
Trevor Burton

David Bushman
Adrian Butcher
Colin Butcher
Kenneth D Butler Jr.
Nicholas Butta
Gary Butterworth
Matt Buxton
Joachim Buyle
David Bylander
Paddy Byrne
Ken Bywater
Alain Caillet
Patrick Cain
Xavier Caine
Kevin Cameron
Jason Camlic
Iain Campbell
Matthew Campbell
Paul Cannon
Jonny Cantwell
Grégoire Capron
Sean Carey
Matthew Cariseo
Michael Carley
Juan Carlos Rubio Román
Adrian Carnally
Roberto Carnevali
Jonathan Carpenter
Christopher Carr
Wes Carr
Tom Carrell
Nicolas Carreno
Brad Carson
Stuart Carson
Russell Carter
Jérémy Cartier
William Carver
Christopher Case
Roger Castela
António Castelo
Jesús Caudeli
Stephen Caulfield
George Caveney
Maxime Cerdan
Kenneth Cervantes

Teddy Chadd
Nicholas Chainey
Partha Chakraborty
Reyansh Chakraborty
Austin Chamberlain
Phillip Chambers
Bing Chandler
Kasey Chang
PHIL Channer
Becky Chantry
Russell Chapman
Etienne Chartrand
Flak Chen
Stephen Chester
Ralph Chilton
George Chimples
George Ching
Andrew Chorney
Reuben Chown
Eivin Landt Christensen
Christophe Christiaens
Kim Christiansen
Dan Christie
Darren Christopher
Hin Chua
Don Church
John Churchill
David Cini
David Clancy
Philip Clare
Daniel Clark
Gavin Clark
James Leonard Clark
Thomas Clark
Jason Clarke
Sébastien Clavier
Carl Clemens
Benjamin Clerkin
Matt Cloke
Matt Cloke
Wayne Lee Cluett
Stephen Cobden
Alistair Coleman
Linda Coles
Andy Collins

John Collins
Leigh Collins
Luke Collinson
Greg Collyer
Paul Comis
David Conner
Patrick Conner
Sean Connick
Seb Z. Connolly
Matthew Connor
Doug Constable
Mark Conway
Chris Cook
James Cook
Matthew Cook
Alan Cooke
Graham Cooke
Richard Cooke
Anthony Cooper
Duke Cooper
Jeff Cooper
John Copeland
Joshua Copeland
Nick Coppin
Hugo Coqueret
Ray Cornelison
Will Cornish
Alex Coronado
Joel Corrente
Dan Couchman
Jonathon Coughlin
Tony Coult
Adam Cousins
Stephen William Coutts
Tim Cowlam
Paul Cox
Edmund Coyne
Sebastian Craenen
Andrew Craig
James Crate
Simon Craven
Lloyd Crawford
Kelvin Crofskey
Mark Crofskey
Robert Croft

Killian Cronin
Noel Cronin
Ben Cross
Cal Cross
Mary R. Crumpton
Hector Cruz
Raymond Cruz
Tony Cullen
Andrew Cumming
Gordon Cumming
Jason Cumming
Tim D'hoker
Marek Dabrowski
Jorge Dager
Lars Dahl-Hansen
Colin Dahle
Andrew Daly
Christopher Daly
Andrew Dalziel
Avisek Das
Ottar Davidsen
Oliver Davidson
Tom Davidson
Alan Davies
Michael Davies
Andrew Davis
Spencer Davis
Jerome Dawson
Marcel de Groot
René de Jonge
Léon de Perthuis
Victor de Rochebouët
Manuel de Sales
Lad Decker
Jos Decoster
Paul Deegan
Alexander Deese
Fernando Del Amo
Bob Delaney
Dario Dellavalle
Sylvain Deloire
Thomas DeLoughery
Chris DeMattia
Alex "ChaCha" Denton
Christopher Derrick

Alan Devine
Sean Devitt
Dennis Di Franco
Scott Diamond
Stuart Dickson
Tim Dill
Tom Dimaline
Dan Dimitrescu
Alexis Dimocas
Kyle Dinic
Etienne Dion-Bergeron
Kevin Dipper
Peter Dixon
Nick Dobbin
Chris Dobbs
Joshua Dodson
Chuck Dolan
Jochen Dornwald
Michael Doscher
Will Dossel
Hugh Douglas
Shawn Douglas
Patrick Dowling
Vali Dragnuta
Timothy Driscoll
Matthias Droste
David Drouet
Simon Drury
Alice Dryden
Kenny Dubnick
Chris Ducklin
Martyn Duckmanton
Ole-Morten Duesund
Will Duffield
Evan Duncan
William Duncan
David Dunlea
David Dunwoody
Matt Dupuy
Scott Durham
Matt Durrant
Kevin J.S. Duska Jr.
Veratrin Dye
Stephen Earley
Rory Easson Baxter

Gregg Easterbrook
Derek Ebbrell
Simon Eccles
Tim Edmunds
Gareth Edwards
Gavin Edwards
Philip Edwards
Jörgen Efverman
Brian Eilbes
Felix Einarsson
Johan Ek
Erin Eldridge
Dan Ellingsen
Craig Ellis
Michael Ellis
Gölkem Elmas
Andy Elms
Louis Emmett
Frank Endrullat
Dan Ene
Eric Engelmann
Niklas Engler
Mike Enos
Nate Ericson
Neil Ericson
Gareth Evans
Rhys Evans
RJ Evans
Tom Evans
Mike Everest
David Evetts
Morgan Ewing
Matthew Exley
Patrick Roman Fabri
Harley Faggetter
Gary Fairley
Thomas Falmbigl
HJ Fantaskis
Byron Farrow
Matt Farrow
Michael Fatsi
Alistair Feltham
Santiago Fernandez
Andrew Ferris
Luis Figarella

Filippos Filippakopoulos

Dirk Firestorm

Florian Fischer

Michael Fischer

Patrick Fischer

Keith Fisher

Trevor Fitchett

Will Flannigan

Donal Flattery

Steve Fleming

Ian Fletcher

Paul Forrest

Michael Forster

Tony Foster

Bryan Fox

James Fox

Jeremy Fox

Mr F Fox

Quillon Fox

Simon Fox

Randy Frahm

Myles Francis

Simon Franklin

Edi Franković

Jeffrey Frankston

Louise Fraser

Isaac Frazier

Stefan Frei

Hope Fremmerlid

Jaden Fremmerlid

Andrew Frewer

Robert Friend

Peter Frogley

Jerry Frost

Sam Frost

Richard Frow

Eoin FRSA

Jean Francois Fulconis

Ben Fuller

Jon Fuller

James A Galea

Sqn Ldr A. P. Galea (Rtd)

Chase Gallagher

Sean Gallagher

Chris Gallon

Mark Gamble

Liam Gannaway

Benjamin Garcia

Bob Gardere

Bob Gardere

Daniel Gardner

André Gaudin

Geoff

Frank Gebelein

Roderick Geddes

Geoff

GGM CE AFJ

Spencer George

Mark Gethings

GGM CE AFJ

Sebastian Giardina

Andy Gibson

Lee Gibson

Steve Gifford

Laurence Giles

Dan Gilson

Luis Giroldi

Vincent Giroux

Jonathan Gitlin

Richard Gleitsmann

Florian Glock

Damien Gobry

Andy Godfrey

Mark Godfrey

Jonathan Gold

Ed Golden

John Gomez

Steven Gonzales

Manuel Gonzalez

Howard Goodall

Tom Goodfellow

Gary Gordge

James Gordon

Tom Gordon

GrafEisen

Oscar Gothberg

Antony Grabham

Mike Grabski

GrafEisen

Terry Graham

Andrew Granville Smith

Jarvis Greatwolf

David Green

Matt Green

John Greenstine

Justin Greer

Jason Gregory

Dominic Griffith

John Griffiths

Matt Griffiths

Mike Griffiths

Richard Griffiths

Mike Griggs

Berrie Grob-Pronk

Henrik Grønbæk

Edward Gronenthal

Kaji Guest

Tom Guest

Thibaut "TIB" Guilbert

Aarron Gull

Paul Gunn

Trevor Gunnell

Connor Gunnin

Per O Gustavsson

Erlend Guttu

Roxy Gwynn

David Haas

Oliver Habibi

Serge Hadfield

Robin Häggblom

Pål Høiland Hagtvet

Paul Hale

Benjamin Hall

Charlie Hall

Kelly Hall

Neil Hall

Phil Hall

Ryan Hall

Sam Hall

Seb Halsall

David Halsey

Mark Hamilton

Paul Hamilton

Stephen Hamilton

Paul Hand

Mark Hanneman

Nathaniel Hansen

Martin Harach

Shane Hardie

Alexander Harding

James P Harker

Doug Harley

Neil Harper

Roger Harr

Christopher Harris

Oliver Harris

David Hart

Logan Hartke

David Harvey

John Harvey

Timothy Harvey

Edward Hass

Andreas Hauck

Blair Haworth

Eoghan Hayes

Steve Hayes

Matthew Hayward

Abhinaba Hazra

Oliver Hebson

Keith Hector

Theo Heddlesten

Jan Hedström

Malte Heiden

Falk Heiland

Jim Heimerl

Conrad Heiney

Christopher Heiny

Olav Heirman

Bruce Held

Russell Hellyer

Jonathan Helm

Les Hemmings

Brian Henchey

Brian Hendrickson

Ian Hepher

Joachim D. Heppner

Quentin herbette

Russ Herrold

Douglas Hewstone

Jeremy Hibbert

Graham Hicks

John Higgins

Sean Higgins

Dan Higgott

Mark Hilborne

Mark Hill

Paul Hill

Michael Hind

Brian Hitch

Ivan Ho

Jeffrey Ho

Gerald Hoag

James Hodden

Bob Hodges

Martin Hodges

Neil Hodgins

Martin Hodgson

Dr. Holger J. Hoffmann

Wulf Höflich

Stuart Holcroft

Chris Holland

John Hollander

Henry Hollick

Christopher Hollingworth

Chris Hollinrake

Geoffrey Holloway

John Hilton Holloway

Brett Holman

Al Holmes

John R Holmes

Mark Holmes

Matthew Holmes

Neil Holmes

Andrew Honeywill

Radek Honzák

Alasdair Hood

Chris Hopkins

Edward Hopkins

Kyle Horan

Desert Hornet

Mick Horsley

Chad Horton

Damien Horton

Jim Howat

Steven Howell

Kristian Howgate

Martin Huberman

Helen Hubert

Keil Hubert

Richard Huffman

David Hughes

Matthew Hughes

Michael Hughes

Shaun Hughes

Jussi Hulkkonen

Silas Humphreys

Anthony Hunt

Edward Hunt

Jack Hurley

Charles Hurst

Patrik Hurtig

Windo Hutabarat

Stephen Hynes

Omar Khaleel Ibrahim

Simon Iddon

James Ironside

Luke Irwin

Khairil Isa

Kevin Isaac

David Isby

John Isham

Hiroki Ishida

Alan Jackson

Dana Jackson

Danny Jackson

David Jackson

Malachi Jackson

Matthew Jackson

Richard Jackson

Hans Jacobs

Zoey Amelia Jacobs

Peter Jakobsson

Simon Jakubowski

Brian D. James

Chris James

Simon James

Jim Jamieson

Paul Jankura

Martynas Janonis

Mark Jarrold

Joel Järvi

Richard Jarvis

Robert Javan

Leigh Jeffs

Leigh & Lewes Jeffs

Alan Jenkins

Jules Jenkins

Stuart Jenkins

Gordon Jennings

Daryl Jewell

Nimish Jha

Ken Johns

Barrett Johnson

Darren Johnson

Edward Johnson

Matthew Johnson

David Johnston

Richard Johnston

Stephen Jolly

Evan Jones

Gareth Jones

Gareth Jones

Jack Jones

Jacob Jones

Justin Jones

Ken Jones

Matthew Jones

Rhodri Jones

Sam Jones

Jymmys

Klina Jordan

Michael Jung

Eric Junker

Neil Jury

Mark Juster

Jymmys

Hinrich Kaestner

Juha Käki

Teemu Kalijärvi

Peter Källviks

Dimitrios Kalogiannis

Lincoln Quinn Kane

Ron Kaplan

Gabriel Karaboulad

Erwin Karincic

Ben Karsian

Tajeshwar Kaul

Rémi Kaupp

Philip Kaye

Andy Keay

Ronan Keenan

Graham Kehily

Jeremy Kelleher

Chris Kelley

James J Kelly

Patrick Kelly

Sean Kelly

David Kemp

Markus Kempke

Paul Kennard

Al Kenny

Andy Kenyon

Glen Keywood

Siddharth Khare

Rich Kidd

Steve Kiener

Dan Kieran

Ian Kilburn

Matt Kilcast

Neil Kiley

Martin Killick

Stephen Kilvington

Connor King

Michael King

R King

Richard King

Russell King

Sheppa King

Stef King

Forrest King-Mathews

Stephan Kippe

Ken Kirgan

Aaron Kirsch

Shutaro Kitagawa

David Kitson

Veli-Pekka Kivimäki

Mike Kluth

Arthur Kluvo

Justin Knapp

Gary Knell

Michael Knight

David Knill

Chris Knowles-Vollentine

Bill Kohn

James Kohn

Ole Konstad

Kev Kopec

Jason Koval

Erik Krane

Vladimir Krapp

Jannis Kraum

Aidan Krier

Yeremia Krisnanto

Matthew Kroening

Robert Krom

Chad Kuboviak

Vikram Kumar

Joe Kunzler

Jeremy S. Kuris

Øyvind Kvangardsnes

Joachim L.

Ralph Lachmann

Nick Lake

Karl Lakner

Mark Lambe

Paul Lammond

Scott H Landry

Robin Landy

Clive Lane

Dr. Daniel A. Lane

Andreas Larsson

Kristian Laskey

Hadrien Lavaux

Raymond Law

Michael Layhe

Nick Le Huray

Dave LeBlanc

Patrick Lee

Mat Leeming

Jari Lehtonen

Rodney Lelah

Gary Leonard

James LePage

Marc Levy

Dawson Lewandoski

Daniel Lewis

Dave Lewis

Richard Lewis

Jerry Lindbergh

Carl Lindgreen

Fredrik Lindholm

Nick Line

John Lingford

Robert Linquist

Philip Liptrot

Justin List

Nick Livingstone

Per Ljung

Nigel Llewellyn-Smith

Alexander Lloyd

Greg Lloyd

Chun "Herman" Lo

Matthias Löbach

Andrew Local

Robert Locke

Jonathan Locker

Landon Locker

Henry Lockett

Daniel Lockhart

Juan Lopez

Aarni Louhio

Mikko Louhio

Thomas Lovegrove

Scott Lowe

Krzysztof Łoza

Bob Lucas

Jason Lucy

Karl Ludvigsen

Robert H. Lund

Ulf Lundberg

Jon Lyons

Sonu M

Ke Ma

Andreas Mac Mahon

Filipe Machado

Hugh Mackay

Andy MacKinnon

Colin MacLaren

Douglas MacLeod

Richard Macpherson Barrett

John Macus

Shreyas Madabushi

Darren Maggs

John Maher

Devraj Mahida

James Mahon

Stefan Maier

Rob Main

Anshuman Mainkar

Simran Majumdar

Christopher Malany

Ian Malcolm

Richard Maley

Ben Malin

Kyle Mallinak

Eran Malloch

Jake Malmgren

Peter Malyon - for Murray Malyon

Christopher "Chris" Man

James Manning

Phil Manning

Scott Mansell

Alistair Manson

Sama Mara

Frank Marasigan

Anthony Marcella

Matthew Markezin-Press

Randall Markgraf

Matej Marko

Frank Markus

Charlie Markwick

Antony Marlow

Ben Marsden

Thomas Marsden

Peter Marshall

John Martin

Justin Niall Martin

Neil Martin

Paul Martin

Jesús Martín Sánchez

Jack Mason

John Mason

Mark Masters

Stephen Mather

Steve Mathews

Duncan Mathieson

William Matzen

James Maugham

Jim Maurer

Mike Maurer

Geoffrey Mawson

Dawood Mayet

Scott Maynard

Sam Maywood

John D McBean

Ian McCabe

Marie McClellan

Michael McClellan

Ryan McCloskey

Connor McConville

James McCormack

Luke McCormack

Ian McCoy

Peter McCrudden

Eamon McCue

Alex McCurdy

Eoin McDonnell

Ian McDougall

George McGhie

Campbell McGill

Kevin McGlynn

Michael McIntosh

Hubert McIntyre

Bailey Mckay

John McLaughlin

Alasdair McLay

Tom McMahon

Andrew McMullon

David McNay

James McNeill

Keith McNeill

Aidan McNelis

Stephen McParlin

Gordon McRae

Jamie McTrusty

Keith Meachem

V Mehos

Dhruv Mehra

Aaron Mehta

Dominic Melville

Prashant Menon

Scott Messinger

Thomas Metke

Stephen Meyer

Andy Miles

Glenn Millam

Adin Miller

Anthony Miller

CJ Miller
Evan Miller
Jeffrey Miller
Robert Miller
Thomas Miller
Jacob Mills
John Mills
Jason B Miner
Justin Miner
Philip Mingle
Paul Minihan
John Miniter
Moellus
Jim Mitchell
Peter Mitchell
John Mitchinson
Girish Mithran
Kyle Mizokami
Pino Modola
Moellus
Duncan Moir
Alastair Monk
Lora Monsees
Martin Montreuil
Jim Monty
Tom Moody-Stuart
Patrick Moore
Brett Morgan
Eugene "Georgio-Scorchio" Moriarty
Yves Morier
Masafumi Moritake
Andreas Morlok
Mark Morris
Derek Morrison
James Morrison
Matthew Morrisroe
Emile Morrissette
Gabriel Moshenska
Benjamin Moss
Richard Moulds
Andrew Mount
Miguel Mourato
Shaun Mower
Scott Mowry

Mark Moxley-Knapp
Korbinian Mrasek
Craig Mudd
Nathan Mueller
Cedric R. Muentener
Aaron Mulder
Holger Müller
Bruce Mulraney
Al Muncey
Antony Murphy
Colin Murphy
Elliott Murphy
Lorne Murphy
Paul Murphy
Richard Murphy
Simon Murphy
CAPT Thomas "Bones" Murphy, USN
Matt Mutt
Paul Mutter
Cedric Mutzig
Michael Mykytyn
Sai Mynampati
Yasuo Nagaoka
Adam Nagel
Jai Nair
Samuel Nall
Shashanka Nanda
Phil Naranjo
Richard Nash
Carlo Navato
Aamar Nawaz
Rahul Nayer
Thomas Nealon
Maxwell Neely-Cohen
Leonardo Daniel "NITRO" Nemec
Daniel Nepomechie
Daniel Newberger
Anne Nguyen
Hung Nguyen
Steve Nichol
Tom Nicholls
James Nicol
Derek Nielsen
Jan Niemczyk

Philipp Maria Niemeier
Timo Niemelä
Andrew Niemyer
Juergen Nieveler
Matt Nightingale
Richard Niland
Johan Nilsson
Max Nippard
Leo Nishihata
Kevin Nolan
Richard Nolan
Zoe Nolan
Paul Norris
Barry North
Colin Norton
Hugh Norton-Smith
Andrew Novak
Serge Novak
Lukas Novotny
Zachary Nupponen-Garcia
Vili Nurminen
Rainer Nyberg
Jonas Nygårds
Audun V. Nytrø
Cormac O'Brien
Luke O'Brien
Sean O'Dell
Anthony O'Donnell
Garrett O'Leary
Janine O'Malley
Andrew O'Marah
Mark O'Neill
Patrick O'Neill
Ed O'Rourke
Stephen O'Sullivan
Justin Ober
Sergio Obradors
Mark Ofield
Owen OHara
Isaku Okabe
Anthony Oliver
Darren Olivier
Mol Olsson
Frank Olynyk
Yasuharu Omi

Jun Hong Ong
Jo Organ
Ant Osborne
Steffen Østenstad
Niall Oswald
Richard Oussedik
Ron Owen
Austin Owings
David Oxley
Andrew Oyston
Atul P
Manoj P
Jim Pacheco
David Paisley
Andy Palmer
Dhaivat Pandit
Anandeep Pannu
Rishabh Parekh
Ian Parish
Chris Parker
Denis Parker
Richard Parker
William Parker
Sam Parkin
Lee Parnell
John Parratt
Tom Parry
Ed Parsons
John-Alan Pascoe
Farbod Pasha
Paul Pasveer
Vaughn Patania
Simon Patrick
George Patterson
Alex Pavloff
Martin Payne
Roger Peachey
Lewis Peake
Rodney Pearce
Stephen Peberdy
Tony Pedley
Javier Pedreira «Wicho»
Dr J G Pelham
James Pemberton
Steve Penny

Shawn Penrod

James Pernikoff

Richard Perry

Mark Persad

Francesco Perucca

Nils Petersen

Ben Peterson

Chris Petty

Jonas Pfeiffer

Sven Pfister

Vijay Philip

Mark Phillips

Mike Phillips

Nick Phin

Ruggero Piccoli

Tom Piel

Keeron Pierce

Will Pierce

Mark Piesing

Allen Piper

Thomas Pitts MRAeS

Pedro Planells Vargas

Peter Plassmann

Marcus Polk

Justin Pollard

Chris Pope

Alessandro Porcu

David Porter

John Porter

John Porter

Simon Porter

Michael Potter

Rob Potter

Robert Potter

Donald Potvin

Fardad Pouran

Tero Poutia

Mark Powell

Evan Powles

François Prenot-Guinard

Chris Prenovost

Phillip Edmund Pretorius

Alexander Price

Jennifer Price

Michael Price

Ole Primdahl

Alex Prine

Florian Prischl

Mark Prüße

Oscar Puerto

Pep Pujol Mur

Mihai Ştefan Puşcaş

Conrad Quilty-Harper

Nicolas - Thanks for pulling G's with
me! Radek

Darren Radley

James Gerard Ragavan

Air Commodore Dinesh Raghavan

Scott Ragon

David Rainey

Harish Rajan

Girish Ramakrishnan

María Barba Ramírez

Jeff Ramsey

Georgi Rangelov

Rainer Rapp

Tim Rasmussen

Tommaso Ravaglioli

Rebecca Raven

Rudradeep Ray

Robert Redmond

Simon Rees

Fraser Reid

Miguel Renderos

Adam Reynolds

James Reynolds

Will Reynolds

Anthony Rhoda

José Ribelles Fayos

Brian Rice

Clive Richards

Daniel Richards

Thomas Richards

C Richardson

Dave Richardson

B Craig Richmond

Neil Riddoch

Eric Rieder

Bennett Ring

Adam Ringel

Edward Rippeth

John Rippeth

Aaron Rivamonte

Gabriele Rivera

Chris Roberts

Douglas Roberts

Paul W Roberts

Peder Roberts

William Roberts

Andrew Robertson

Ben Robertson

Andy Robins

Tim Robinson

Keith Rockhold

Toni Rodrigo

Pedro Rodrigues

Tobias Roehrig

Ivan Rogers

Scott Rogers

Stephen Rogers

Pieter Röhling

Rotopenguin

Fredrik Rönnblad

Paul Rosencrantz

Darrell Rosenstein

Scott Rosenthal

Thomas Ross

Tom Ross

Ari Rotonen

Rotopenguin

Jordan Rouse

John Routledge

Roberto Roux

Arthur Row

Philip Rowles

Abhishek Roy Bardhan

Timothy Ruggieri

Bill Rumbach

Christopher Ruse

Kevin Russell

David Ryan

Eoin Ryan

Michał Rydzy

Roland Sadowski

Mario David Saitto

Jay Sala

William Salley

David Sallows

Nikolas Sambado

Chakrit Samithinan

Travis Sampiero

José Sánchez

Edward Sandstig

Víctor Sanguino

Tom Sanor

Siraphob Santironnarong

Akira Santos Stefan

Abhilash Sarhadi

Shiv Sastry

Steve Satak

Mark Satchell

Damien Saunders

Mark Saunders

Georgi Savov

Jack Sawford

Phil Sawyer

Prateek Saxena

Peter Scales

Åsa Scharin

William Schauman

Spencer Scherer

Michael Witus "Gix" Schierup

Paul Schifferes

Nathan Schlehlein

Nicholas Schmidt

Samuel Schmitt

Gabriel Schonfeld

Cole Schoolland

Nate Schutte

Christopher Sciberras

Andrew J Scott

Bradley Scott

Doug Scott

Robert Scott

Kelvyn Scrupps

Dan Seagrave

David Seall

Rory Semple

Ian Servin

Xavier Seuillot

Gaurang Shah
Dan Sharp
Michael Sharp
Peter Sharpe
Siarhei Shauchenka
Alan Shaw
Chris Shaw
Loren Shaw
Robert Shaw
Jayson Shenk
Mark Shepherdson
Sam Sheppeck
Keith Shiban
John R. Shibley
Alistair Shields
Daniel Shingleton
Kenta Shinjo
John Shufflebotham
Andreas Sikkema
Richard Sikorski
Eric Silski
Ian Silvester
Michael Silvius
Matt Simpson
Andrew Sinclair
Louis Rundio Sinclair
Angad Singh
Bhanu Singh
Kartikeya Singh
Rajnish Singh
Udaybhanu Sinh
Michael Sinkowitsch
Phil Sivills
Slenderhose
Ulf Sjöström
Judy Skidmore
Gordon Skillen
Luke Skipper
Anthony Slack
Adam Slater
Guillaume Slee
Slenderhose
Howard Slutsken
Thomas Smail
Alan Smith

Carl Smith
Colin M. Smith
David Smith
David Smith
Gary Smith
Harold Smith
James Smith
Jayden Smith
Martin Smith
Max Smith
Patrick Smith
Pete Smith
Richard Smith
Ron Smith
Samuel Smith
Tom Smith
Tyler Smith
J. Smith Sr
Harrison Smithey
Robert Snarr
Harmen Snel
John Snodgrass
Michael Snow
Scott Snowden
Phillip Sobel
Tomas Sogndal
Parsifal Solomon
Stefan Solomon
Zachary Solomon
Alastair Somerville
Neil Sood
Jaakko Sotkasiira
Pedro Soto
Nikky Southerland
James Spahn
George & Theodore Spahr
Deni Spasovski
Damian Specht
Adam Spink
Asher Spivak
Dave Spring
Noah Sprinkle
Krishnan Srinivasan
Matthew St. Onge
Jonathan Stace

Dominic Stadden
Frank Stadmeyer
Phil Standring
Edward Stanley
Glenn Stanley
Kirk Stant
Eamon Stapleton
Harry Staruk
Casey Stedman
James Steele
Eric Stepans
Simon Stephen
Shane Steptoe
Mark Stergios
Scott Stergios
Edward Stern
Darren Steven
John Steven
Sally Stevens
John Stevenson
Ben Stewart-Reed
Matthew Stibbe
Matthias Stohr-Niklas
Simon S Stokes
Stuart Stones
Michael Stoodley
Scott Stopher
Al Storer
Joseph Stronach
Brian M Stroud
Robert Stroud
Ian Stubbs
Jon Stump
James Sturgill
Terry Suitor
Kasperi Summanen
Paul Sumner
Markus Suojoki
Chris Suslowicz
Richard Sutcliffe
Mark Sutherland
Erich Swafford
Robert Sweetnam
Mark Symons
Ian Synge

Richard Tanner
Andy Tarbard
Russell Tarr
Michael Tawton
David Taylor
Paul Taylor
Pete Taylor
Rich Taylor
Simon Taylor
Carey Taylor-Forbes
Nathan Tehrani
Thomas Terashima
Shane Thacker
Advent the318
Amdi Thellemann
Clive Thomas
Nick Thomas
Xavier Thomas
David Thombs
Ben Thompson
Patrick Thompson
William James Thompson
Andrew Thomson
David Thornton
Anders Thorsen
Tan Tian Cai
Paxton Ticoulet
Tuomo Tilman
Ted Timmons
Aquiles Tinoco-Moreno
Ian Titley
Jonas Tjernlund
Michael Todd
Hiroshi Tokutomi
Nicholas Tones
Bernard Toogood
Don Tordilla
Stephen Tosh
Phil Toste
Nils Toudal
Piers Townley
J. R. Tracy
Bob Trickey
Robert Trimble
Adrian H Troughton

Wayne Trowsdale

Twats

David Tsui

Chris Tucci

Henry Tun

Juho Tuomisto

Travis Turk

Andrew Turnbull

Mary Frances Turnbull

Steven Turner

Twats

Mike Tycoles

Ben Tymens

Rich Tysoe

Arne Ubelhor

Alex Ucci

Benjamin Umstead

Nicolas Untz

Robert Upton

Gabriel Uriarte

Kedar Vaidya

Paul Valenza-Troubat

Andries van Bergen

Rienco van der Mooren

Frank van Dijk

Dan Van Heeswyk

Jurriaan van Ingen

Arjan Van Krimpen

Jonne van Lunteren

Pieter Jan van Neerbos

Elton van Steenbergen

Denis Vandermeulen

Kurt Vanhollebeke

Phil Vaughan

Tomas Verbaitis

Carlos Vergara

Eric Vernier

Anthony Verow

Gregory Viallard

Kasper Viita

José Vizcaíno

Ian Vobe

Stefan Vogel

William Wagner

David Wahl

Nigel Walker

Thomas Walker

Alfred Wallace

John Wallace

Andrew Wallbank

Bernard Walsh plus Howard Martin

Carl Waltenberg

Brandon Walters

Daniel Waltimire

Liam Walton

Geoffrey Wang

Thomas David Ward

John Waterhouse

John A Watson

Adam Watt

Martin Weaver

Paul Matthew Webb

Patrick Weber

Thomas 'Spart' Weber

William Weeks

Wouter Weersink

Samuel Weidenbach

Michael Weinold

Rich Weite

Julius Welby

David Weller

Mike Welsh

Tim West

Tony Weston

Tae Whang

Simon Wharton

Clayton Wheeler

Stephen Whelan

Damian Whitby

Mike Whitcombe

Gary White

Pete White

Rowland White

Christopher Whitehouse

John Whitehouse

Jakob Whitfield

Ross Whithorn

Timothy Whittaker

Matthew Whittington

Gert-Jan Wiarda

Thomas Wigley

Eric Wijngaarden

Darryl Wilbraham

David Wilcox

Joe Wilding

Helge Wilker

Ian Wilkins

Mike Wilkins

Simon Willatts

Martin Willcox

George Willett

Adrian Williams

Alistair Williams

Chris Williams

Daniel Williams

Dominic Williams

Tom Williams

Pete Williamson

Grahame Wills

John Wilson

Mathew & Robert Wilson

Peter Wilson

Rick Wilson

Derrick Wiltshire

James Winchester

Christopher Windsor

James Windsor

Michael Winter

Sam Wise

Eva Wiseman

Nadia Wiseman

Gregory Witpen

Marc Witsel

Thomas Wittenburg

Marc Wolff

Brian Won

Marcus Wong

Steven Wong

Andy Wood

Larry Wood

Nick Wood

Stephen Wood

Larry Woodall

Ken Woodruff

Neil Woodward

Steve Woolmington

Paul Wren

George Wright

Julianne Wright

Matt Wright

Leon Wu

Roger Wu

Andrew Wyatt

Jason Xu

Joel Yang

Michael Yanikoski

Chris Yates

Shayne Yates

Andri Yatim

Ricardo Ybáñez Harguindey

Barry Charles Yodzis

Omer Yousuf

Masanori Yusa

James Zapf

C Zwinkels

Γιάννης Μικές

Олександр Р

伊知郎 吉田